Germ Wars

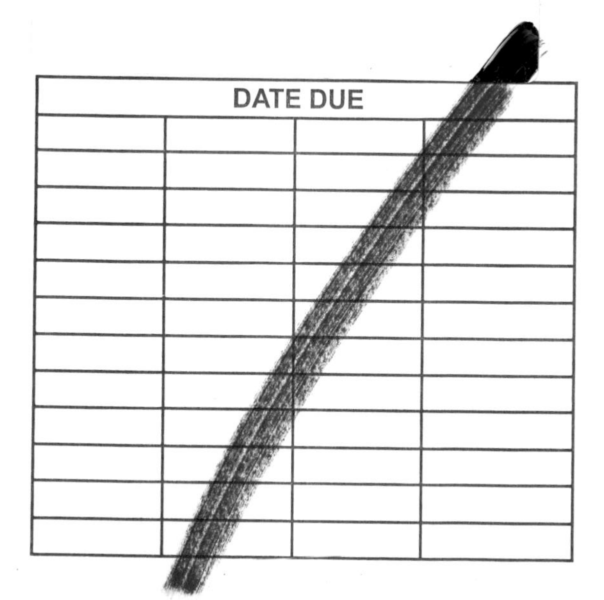

CRITICAL ENVIRONMENTS: NATURE, SCIENCE, AND POLITICS

Edited by Julie Guthman, Jake Kosek, and Rebecca Lave

The Critical Environments series publishes books that explore the political forms of life and the ecologies that emerge from histories of capitalism, militarism, racism, colonialism, and more.

1. *Flame and Fortune in the American West: Urban Development, Environmental Change, and the Great Oakland Hills Fire*, by Gregory L. Simon
2. *Germ Wars: The Politics of Microbes and America's Landscape of Fear*, by Melanie Armstrong

Germ Wars

The Politics of Microbes and America's Landscape of Fear

Melanie Armstrong

UNIVERSITY OF CALIFORNIA PRESS

University of California Press, one of the most distinguished university presses in the United States, enriches lives around the world by advancing scholarship in the humanities, social sciences, and natural sciences. Its activities are supported by the UC Press Foundation and by philanthropic contributions from individuals and institutions. For more information, visit www.ucpress.edu.

University of California Press
Oakland, California

Library of Congress Cataloging-in-Publication Data

Names: Armstrong, Melanie, 1977– author.
Title: Germ wars : the politics of microbes and America's landscape of fear / Melanie Armstrong.
Description: Oakland, California : University of California Press, [2017] | Includes bibliographical references and index.
Identifiers: LCCN 2016037674 (print) | LCCN 2016039496 (ebook) | ISBN 9780520292765 (cloth : alk. paper) | ISBN 9780520292772 (pbk. : alk. paper) | ISBN 9780520966147 (epub)
Subjects: LCSH: Bioterrorism—United States—Prevention. | Bioterrorism—United States—Psychological aspects. | Biopolitics.
Classification: LCC HV6433.35 .A76 2017 (print) | LCC HV6433.35 (ebook) | DDC 363.325/35610973—dc23
LC record available at https://lccn.loc.gov/2016037674

Manufactured in the United States of America

25 24 23 22 21 20 19 18 17
10 9 8 7 6 5 4 3 2 1

Contents

Introduction

Political Ecologies of Bioterror

I was working in the visitor center at Arches National Park in the fall of 2001 when we received a report of suspicious behavior in the park. The reporting party said a man who appeared to be "Middle Eastern" was dusting a white powdery substance around the base of Delicate Arch, the sandstone span venerated on the Utah state license plate and in the pages of sporting goods catalogs. We put out a call to a law enforcement ranger, and within minutes a cavalcade of park rangers, county sheriffs, city police, and firefighters rushed into the park with lights spinning and sirens blaring. From the front gate, we watched the cars streak by and listened to the tight-jawed radio traffic, imagining the spectacle that was shattering the serene desert landscape a few miles to the north. We pictured bodies clad in Kevlar vests and thigh-high fire boots clunking up the mile-and-a-half-long trail. We envisioned hikers in Patagonia polos gathering near the arch for a late afternoon meditation or to munch on granola bars and imagined the look on their faces when a SWAT team pounded over the horizon brandishing batons and body shields. Later we speculated about how long it had taken the man with the dark complexion to convince the rangers that the powder on his hands was not anthrax, but the ashes of his dead father. He had come to this place to ritualistically honor and remember someone he loved.

Imagining the emotions of that confrontation—confusion, panic, anger, grief, insult, and fear—awakened in me an understanding that the terrorist events in the United States earlier that year were changing the world. More than new laws or renewed patriotism, these changes were manifest in the ways people interacted with each other and how they read the bodies and behaviors of other human beings. Three months earlier, the report might have attracted the attention of one ranger, who would have reminded the man that it violates the Code of Federal Regulations to spread ashes in a national park. Onlookers would have inquired what he was doing and perhaps asked questions about his dead father and why he'd selected this scenic resting place. But white powder had a new meaning after the attacks of September 11, 2001, and when carried by a person of color, the substance provoked not feelings of respect for the dead but fear of the living. Even a remote desert landscape near Moab, Utah (population 5,000), was remade by the new materialisms of a war on terror. Small-town cops and national park rangers were enlisted in a nationwide project to secure every space, every citizen, and every icon of American life. No landscape was exempt from the security practices implemented around the country over the next few years.

In the national park, we covered our electrical outlets, locked doors to the restrooms, and subjected our volunteers to formal background investigations. Two years later, in Yellowstone National Park, I watched a bomb squad from Denver take a briefcase we had found in the visitor center into a meadow and blow it up. The case was likely a tourist's misplaced collection of maps and guidebooks, but by this time we were trained to treat it as a potential weapon designed to wreak havoc at a public park. By this time, I recognized that I was participating fully in the production of this new security regime. Citizens like me submitted to airport security, and workers like me radioed in suspicious-looking briefcases, creating through our actions the threatening world we expected to see.

Startled though I was by the level of response I saw at Arches, I had spent enough time working in national parks to realize that people's relationships with natural landscapes are complex and deeply rooted in notions of power (and powerlessness), reverence, and nationhood. Over time, people have produced ideas of nature as a wild place, moral compass, or selective breeder, enacting a cultural politics of nature through the daily management of place.[1] Nature is also a site of political struggle where both meaning and materiality are contested. The events at Delicate Arch showed, for example, that the use of nature by people of color is suspect. These naturalized assumptions about race are being remade in the twenty-first century. The ways we engage with nature forge new material forms, and the overlapping struggles between nature and culture, object and idea, constitute the world. In 2001, the management of nature became part of a larger project to create "homeland security," bringing with it naturalized political struggles over race and class. When the actions of a person of color in a place claimed by white tourists require violent policing, this new national-security politics collides with a cultural politics of nature, provoking consideration of the social effects of how we engage with people and landscapes. This management of nature is also the governance of citizens. As I witnessed in the emergency intervention at Arches, the new terms of citizenship constructed in the interests of national security are creating mechanisms for producing modern natures, even as these new natures create new futures for modern citizens.

There were other suspect elements in the scenario at Delicate Arch. White powder, particularly in the hands of a person of color, connects living organisms to political violence and represents a new form of nature that is unseen, deadly, and subject to human manipulation. White powder is the visage of bioterrorism in the twenty-first century, materializing the possibility that microbes can be managed by humans to inflict harm. It stands in for broader cultural fears of the unpredictability and intentionality of nature, even as it constitutes killer natures in a form that can be powdered, packaged, and sent through the mail. White powder symbolizes the collapsing of dualisms: living organisms have been technologically recreated in forms that could not exist without human intervention, yet they resist submission to human will. These are the new forms of nature that produce and are produced by the modern security state.

Bioterrorism has emerged as a prominent fear through the cultural production of microbial nature alongside changing practices of warfare. Revolutions in biological science in the last hundred years have created a vast microscopic world, and in this same era we have watched the rise of a global war on terror. Though these movements appear to have emerged separately, they are deeply entwined. The science of modern warfare is the science of nature, pursuing new knowledge of life and death. In turn, this new knowledge of nature has made possible new mechanisms of war.

Microbes have been created through science practices that reconstitute their natures and change their relationship to the human world. Science has expanded the possibilities of microbial life in ways that produce fear and challenge our anthropocentric understanding of the world. The knowledge that unseen microbes permeate all environments, including the human body, changes how people live in the world, which in turn transforms microbial landscapes. Emergent knowledge of microbial life, produced in modern social contexts, further shapes the interspecies relationship through ever-changing conceptions of intimacy and dependence. Microbiotic communities enter into the human body before birth and cannot be extracted without destroying life. Humans are no longer understood as discrete organisms but are hybrids with their microbes.

Scientists have further shown that homes and built environments once deemed separate from the natural world are suffused with microbial nature. On every surface dwell organisms that might harm human health and life. New fears have coalesced around the intrusion of disease-causing microbes into our homes and lives, and mechanisms have emerged for managing "germs." The history of our work to create and then harness and control germs reveals how biological natures actualize their own formation: the microbe's material form embodies human fears and desires. A biological weapon, therefore, materializes the belief that humans can harness nature to inflict harm for political purposes, along with the desire to control forms of nature that might harm us. The rise of bioterrorism encapsulates emergent social fears around new biological forms of nature and the vulnerability of human life.

The United States government has spent more than $215 billion since 2001 to prepare the nation for bioterrorism. Examining the political context for the rise of bioterrorism as a powerful cultural force exposes a broader shift in how people think about nature, nurturing new fears of microbes that rationalize massive government expenditures and peculiar political interventions. In this book I explicate two key themes in the natural history of bioterrorism. First, the materiality of the microbe as a small, unseen, environmentally specific organism supports a cultural belief in nature as an entity that can be managed according to human desires. Humans have long produced nature in the form of plants and animals that can be managed, and the same mechanisms of control have simultaneously been engaged to manage microbes. Thus the production of microbial natures mirrors the historical productions of nature on the scale of landscape or ecosystem. The political histories of disease have coalesced with modern bioscientific practices to create a collective understanding of bioterrorism cohesive enough to mobilize social action. Second, we have built new institutions of governance for this microbial world, thereby changing what it means to be a human and a citizen. Our social systems look different because they were created to manage the microbe. These politics not only produce new ideologies of security, community, and nationhood, but also create new material worlds. I survey the landscape of bioterrorism—and prod at its formation—to show the material outcomes of bioterrorism in modern life. Regardless of whether there has ever existed a viable bioterrorism threat, the fear that microbes will harm humans is remaking social systems, compelling modern subjects to work to secure their bodies and nations against biological threats.

Enlisting the bioscience industry in the work of national security produces new systems of power predicated on rearticulations of life, death, nature, and disease. The emerging spaces and economies of bioterrorism exemplify how the fusion of genomic, microbiotic, technologized biologies with new forms of warfare brings material change to the lives of people not at war in the traditional sense. These technoscientific extensions of war have militarized the care-giving acts of governance, aligning the objective of governments to maintain a healthy and productive population with the purposes and mechanisms of national security. When biosecurity involves stockpiling vaccines and scanning liquids at airports, the work of protecting the nation touches everyone's life.

The ways in which bioterrorism is produced have consequences for the ways people live in this world of unspecifiable risks. In this introduction, I outline histories of bioterrorism and germs to help define the tensions of the present moment. Next, I present the theoretical underpinnings of the questions I ask about modern society through the examination of bioterrorism. I lay out the terms of biological security as they have been coopted by disciplines and authors, myself included, and chronicle a specific progression of bioterrorism events that articulates with the questions at hand. Finally, I make a case for the consideration of bioterrorism in what Jane Bennett refers to as the "political ecology of things" as a way to account for the vibrant materialities we are making around microbes.[2]

A "TOP THREAT"

"Bioterrorism Seen as Top Threat" was the headline for a February 23, 2007, report from United Press International stating that more than one-third of the ten thousand respondents to a UPI-Zogby International poll identified the fear of biological attack as the number-one health risk facing Americans, outranking the fear of avian flu or HIV/AIDS by more than 10 percentage points.[3] In modern U.S. history, though more than six hundred thousand people have died of HIV/AIDS, only five people have died in bioterrorist acts.[4] In spite of the rarity of biological attacks, everything from government spending to prime-time television drama frames bioterrorism as one of the great threats to American life and society, demanding action in the present to resolve a catastrophe in an imagined future. Biosecurity, the movement to protect the nation from such biological threats, has become a multibillion-dollar industry. Whether or not a "real" biological threat exists, this outpouring of funds is reshaping American communities.[5]

Bioterrorism is a form of terrorism in which living organisms are technically manipulated to inflict harm on a population and instill fear. The fear of biology—a betrayal by nature—has a complex history. Advances in science and technology have expanded the arsenal of the terrorist, allowing us to imagine that anyone from a religious extremist to an adolescent science geek might be building biological weapons in the basement. Even the spectacular images of the World Trade Center collapse in 2001 fell short of framing the real fears of the new millennium. In nearly perfect theatrics, the opening scene of airplanes, skyscrapers, fire, and explosions shifted to the insidious mystery of a white powder transported through the U.S. postal system. The anthrax scare introduced the drama of terrorism into the lives of everyone who opened mail or inhaled air. President George W. Bush called this the "second wave of terrorism": this language marked bioterror as the threat of the future and enabled swift government action to be directed toward a national plan for biosecurity.[6] The U.S. government began preparing for a future in which the use of biological weapons appeared certain.

Fear of pandemics is not new to the modern era, but human relations to disease have been transformed over the past century. Biomedical innovations such as vaccines and antibiotics have for the most part fallen short of early promises that they could eradicate disease from the planet. The AIDS crisis of the 1980s rekindled fears of infectious diseases, and the inability of the science community to rapidly identify the biological agent of the disease startled and outraged citizens. In the new millennium, cases of severe acute respiratory syndrome (SARS) and H1N1 (swine) flu cast infectious diseases as a threat that could spread across the globe, stirring concern that transportation networks would rapidly transmit diseases around the world.

Even the eradication of naturally occurring smallpox in the 1970s created a new sense of vulnerability by drawing attention to the human ability to harness the destructive power of microbes. The suspicion that terrorists and rogue nation-states harbor illicit stocks of the smallpox virus raises fears that the germ will be used to attack a population no longer vaccinated and thus again susceptible to infection. Laboratory research has also raised concern that smallpox could be artificially recreated from its DNA sequence or spliced with genes from other organisms to create a "superbug."[7] The fear of smallpox is no longer the fear of how an organism can destroy human life but of how scientists, politicians, and terrorists can bend a germ to their bidding.

For now the possibilities of biological warfare remain largely imaginary. Only two significant acts of bioterrorism have been recorded in modern U.S. history. Anthrax killed five people in 2001 and sickened at least seventeen others. In 1984, a religious group in Oregon gave 750 people stomachaches by sprinkling salad bars with salmonella cultured in a secret underground laboratory. No one died from those tainted greens, but such incidents have clearly captured the popular imagination. From their sofas, Americans view bioterrorist attacks on primetime television dramas like *Alias* and *24* and feature films like Steven Seagal's *The Patriot.* Such dramas compensate for the absence of real attacks by providing believable enactments of bioterror scenarios. Although microbes are invisible to the human eye and the war they wage does not destroy buildings, such dramatizations render the outcomes visible and therefore more emotionally compelling.

These fictional presentations produce the cultural milieu that gives science and government the authority to act. The dramatization of a bioterrorist attack is typically accompanied by an equally fictional social response and plot twists that create unexpected weaknesses and vulnerabilities. In the 1990s, popular novels like *The Hot Zone* and volumes like *The Coming Plague* fueled public concern over pandemics and produced a terrain on which the biosecurity plans for the new millennium could be mapped.[8] In an anecdote frequently recounted in histories of bioterrorism, the imagination of a novelist led to the first government action to address bioterrorist threats: President Bill Clinton reportedly stayed up all night reading Richard Preston's novel *The Cobra Event* and then ordered copies for every member of his cabinet. Within months of reading the book, Clinton had assembled the country's first bioterrorism taskforce and made the first large-scale funding allocations dedicated specifically to bioterrorism preparedness.[9] This allocation (which preceded the attacks of September 11) exemplifies how material change followed engagement with the cultural imaginary of disease.

The current anxiety over biological attack recalls the climate of fear of nuclear weapons during the Cold War. The United States created a civil defense program as a display of national preparedness, demonstrating, primarily to its own citizens, that the nation would survive a nuclear attack. Cold War drills instructed citizens to "duck and cover," enacting personal rituals to ensure their collective survival. Rehearsing for the future threat was central to public life and a core act of governance.[10] Civil defense demonstrated how the nation would emerge, intact, from the future attack.

During the Cold War, the United States had a clear enemy in the Soviet Union, and the weapon was presumed to be nuclear. In the terrorism crisis, neither the enemy nor the weapon is so readily identified. The logic of bioterrorism preparedness might be called preemptive, as it is based on an uncertain threat. Because a biological threat is perpetually evolving, its nature can never be specified. Preemption "compensates for the absence of an actual event by producing an actual effect in its place."[11] Whether or not a biological threat exists, the work of preparing for bioterrorism is shaping social life. This effect lends certainty to the uncertain threat, enabling citizens to act.[12]

The work of scientists and governments helps citizens discern the uncertain future threats of bioterrorism. Contemporary biosecurity practices exemplify the ways in which the science complex fuels social movements, naturalizing political acts by producing a material world that must be continually managed and controlled. As people make sense of their microbiota, they reevaluate the risks to their life and health, along with their expectations of government to care for their biology. These expectations give rise to mechanisms to gain unprecedented access to citizens' bodies.

Biosecurity also naturalizes the fears that have long sustained the national security state. Some life forms, when manipulated by technologies or reproduced in endless chains of contagion, contain the potential for the destruction of human life. In the case of bioterrorism, the imagined postcatastrophic life is troubled by the assumed endurance of microbial life. Microbes are on this planet—indeed, inside our bodies—for the duration. Biosecurity is not about surviving one event but about building and maintaining immunity against perpetual microbial threats.

Bioterrorism diverges from other classes of terrorism through its direct attack on the human body: no buildings fall in a bioterrorist attack and no bullet holes are left. Only living organisms are susceptible. Controlling one life form to target another creates a particular type of fear. Because a bioweapon is alive, it presents the persistent possibility that the weapon will mutate or escape human control. Bioweapons may also transform human beings or other living organisms into carriers of disease, turning victims into a weapons. The fear that a human could be infected and then knowingly or unknowingly spread a deadly disease creates a unique situation of terror in which bodies are potential weapons, and everyday social rituals create risk.

Scientists have produced abundant knowledge of microbial nature and its threats. The new life sciences are also producing potent sources for remaking systems of governance. In a landscape of fear, people are perpetually vulnerable to unseen threats and therefore seek security. The work of governments to protect citizens is also a form of care. To care for citizens in a world of mobile and mutable microbes, governments must calculate these new microbiological risks and decide how to intervene. The belief that nature can be contained and controlled through governance rationalizes the many social acts that constitute security, and the policing of social behaviors for the sake of biological security shows how the reach of governance has expanded through the production of microbial threats. Through bioterrorism, public systems such as health care and environmental management are brought into the service of national security.

Bioterrorism demands a reconsideration of biopolitics on both the microscopic and the global scale. How do the genomic, microbiotic, interspecies life forms produced in the twenty-first century change our inquiry into the political governance of life? Studies of bioterrorism must seriously consider how Michel Foucault's biopolitics, so thoroughly grounded in the corporeal, are affected by new conceptualizations of human life. Any concept of biopolitics must encompass other living bodies that can literally be absorbed into the human form. As a "visceral protagonist within political encounters," the body is more than just human; it is a permeable collective of living and nonliving matter whose capacity for agency is manifest on many levels, from the competition for resources to social negotiations of power.[13] Human life has fused with microbial life in tenacious and troubling ways. These intimate formations demand a rethinking of the most apparent categories of self and society and result in strange and startling linkages that become vital to understanding modern social existence. Humans and microbes—and microbe-human hybrids—each absorb the logics that have shaped the other, creating new assemblages of nature and culture that are built up and torn apart by the politics of everyday life. The microbe has been made for the modern world: its production is bound up with the rise of neoliberal politics and the nation-state.

GERMS: THE ORIGINS OF LIFE

Contrary to the modern association of germs with disease and death, the term *germ* literally means "origin of life." The earliest forms of life on the planet were microorganisms, and the continued presence of germs on earth ensures the vitality and biological diversity necessary to sustain life. Exploring the origin stories of the germ shows how this complex, even contradictory idea coincides with cultural understandings of life itself and the beginnings of human life and society.

The word *germ* originated in agriculture, referring to the emergence of a new organism inside an existing one. The germ was the being in its rudimentary state, the embryo capable of becoming the organism. This meaning is carried today in the name for the center of the wheat kernel, known as wheat germ. In the mid-nineteenth century, when Louis Pasteur and others studied the hypothesis that life spontaneously generates in decaying meat and fermenting wine, they were seeking the germ of life, the seed that would develop into full-grown maggots or yeasts. To disprove the theory of abiogenesis, scientists had to show that life does not spring from nonliving matter but that life begets life. Pasteur's experiments disproved the theory of spontaneous generation by showing that "germs of microscopic organisms abound in the surface of all objects." He proposed that embryonic microorganisms, invisible to the eye, pervaded the environment and, given the right conditions, could begin to grow and reproduce: "Ferment is an organized being, the germ of which is always present, and the albuminous substance merely serves by its occurrence to nourish the germ and its successive generations."[14] Significantly, Pasteur's origin story also scripted the death of microbes by showing that organisms germinate when conditions are appropriate to propagate life but that these microbes can be destroyed by altering the environments that sustain them. Though he saw the varieties of life on a microscopic scale, in practice Pasteur was concerned with environments. He demonstrated that heat could be used to purify food by destroying the microbes that caused the food to rot.[15] He also proposed that microbes theorized to be the cause of disease in humans could similarly be destroyed by creating an antibiotic environment.

Because he was studying the cause of decay in food, Pasteur's attack on microbes correlated them with rot, disease, and death. Through a semantic specialization, *germ* came to refer to just pathogenic organisms. The term *germs,* with a range of negative connotations, began to stand metonymically for the broader idea of germs as the origin of life.[16]

This change in usage mimicked the cultural shift that accompanied the nineteenth-century study of the microbe: scientists cast germs as the vectors of disease, contaminants in an otherwise safe environment. Microorganisms were targeted as a source of harm, and a word that once referenced life itself now simultaneously encompassed the origins of death and decay, a construction that persists in the present day.

Despite these negative associations, germs play a critical role in human survival. Microbes account for most of the mass of the earth's living matter as well as its biodiversity. As oxygen fixers and recyclers of matter, microbes produce the environment humans enjoy on Earth today. Even as disease bearers, germs stimulate diversity and promote the fitness of particular species. Modern thinking about ecosystems, symbiosis, living matter, or even the Gaia hypothesis incorporates microbes into the understanding of life. As the biologist Lynn Margulis argues,

> Life is an incredibly complex interdependence of matter and energy among millions of species beyond (and within) our own skin. These Earth aliens are our relatives, our ancestors, and part of us. They cycle our matter and bring us water and food. Without "the other" we do not survive. Our symbiotic, interactive, interdependent past is connected through animated waters.[17]

By this conceptualization, *life* means far more than metabolism, growth, and reproduction. Life transcends time and physical boundaries. Life is both alien (that which is not us, not human) and intimately personal (microorganisms living within our own macroorganism.)

Though the microbe is essential to human life, to say that a single microbe is as important as a human would be laughable in many circles.[18] Microbes accrue power in assemblages. Although individual humans perhaps have more potential to alter an environment than individual microbes, given the right conditions, a microbe can greatly expand its reach, multiplying rapidly, changing environments, and bringing destruction to organisms of much larger size. Plagues, poxes, and influenza have caused far more human deaths than technologies like the gun or automobile. Acknowledging the germ as a vital life force with tremendous collective influence opposes the trend of much contemporary science, which looks ever more closely at the chemistry and genetics of the individual organism.

From the earliest portrayals of germs, like William Heath's 1928 illustration of the water of the Thames River as "monster soup," our cultural imagination and media representations have characterized the malice of microbes, extending beliefs deeply into our sociality and align-

FIGURE 1. "A Monster Soup, commonly called Thames Water, being a correct representation of that precious stuff doled out to us." An 1828 etching by William Heath, "dedicated to the London Water Companies," depicts a woman's horror as she looks through a microscope to see the "monsters" living in a drop of water.

ing microbial behavior with broader cultural constructions of good and evil. Because we experience globalization and colonialism, in part, as biological encounters, these forces have shaped how we characterize microbes. Just as today's human ecologies have changed because technologies like air travel transform how people move and interact around the globe, the spread of biota around the globe during the imperialist expansion in the middle of the last millennium also included the spread of microbes. For example, some scholars have attributed the establishment of European power in the New World to the spread of microbes ahead of the arrival of the conquerors, inscribing beliefs about weakness and power on vulnerable bodies (see chapter 1). Microbes can act globally despite their size, as evidenced by pandemics ranging from bubonic plague to swine flu. The rapid reproduction and expansion of microbes ensures that species will survive even as successive generations deplete the resources of one host and move to another. This forward march of disease reads like an invasion. The desire to disempower competitor species, as posited in the 1970s in W. D. Hamilton's spite hypothesis, naturalizes this narrative of conquest, to the extent that zoologists have theorized that nonhuman species may also use microbes to gain advantage over competitors.[19] The study of competition among species and the biological advantages created by pathogens feeds into our popular understandings of microbes as weapons.

The mutability of living pathogens helps to sustain the cultural fear of biological weapons as unpredictable and potentially uncontrollable. Darwin argued that mutation is the origin of species, resulting in the genetic transmission of new traits and the creation of new life forms. Organisms' endless potential for mutation is described by Joseph Masco as a "break with the past that reinvents the future."[20] Through mutations, either natural or engineered by humans, pathogens can develop abilities to spread through air instead of through fluids or to pass from bird to human hosts. Indeed, as the director of the Defense Advanced Research Projects Agency (DARPA) testified before the congressional Committee on Armed Services in 2014 (in remarks shared on DARPA's Facebook page), "Biology is nature's ultimate innovator."[21] Nature's innovation can be hopeful and inspiring, but it can also spread fear and uncertainty about future threats.

In a publication titled *Biotechnology Research in an Age of Terrorism,* the National Research Council underscores the unpredictable nature of microbes but suggests that technology might still harness it: "It may be difficult to engineer a more successful pathogen than those already present in nature that have been perfected by evolution for their niche in life. However, application of the new genetic technologies makes the creation of 'designer diseases' and pathogens with increased military utility more likely."[22] Fear of biological weapons derives in part from the idea that microbes can be manipulated to increase their naturally harmful effects on human bodies. Since 9/11, scientists, policy makers, and citizens have debated how the rise of synthetic biology changes the bioterror risk. While some point to the difficulty of creating microbes that do precisely what humans desire and nothing more, others point to the publication of DNA sequences on the Internet and the creation of a synthetic poliovirus. The human synthesis of a virus, however, does not strip the germ of the power it holds as a living organism. Modern natures, like synthetic viruses, are entangled with the technology and politics that bring them into existence, and as such, their existence requires us to scrutinize the beliefs we have normalized through our discourses of nature and science.

While germs contain perhaps an infinite potential to be technologized, their abundant presence as nature's "kernels of life" also defines their utility for bioterrorism. On its website, the Centers for Disease Control and Prevention (CDC) defines bioterrorism as causing "illness or death [using] agents [that] are typically found in nature."[23] A central fear of bioterrorism arises from the view that microbes occur freely in nature and are therefore perceived to be readily available for malicious use as bioweapons. In a 2015 congressional hearing on bioterrorism preparedness, the former senator Jim Talent testified that the growth of the life sciences has reduced the "barriers to developing a bio-weapon. Disease causing microbes—anthrax is an example—are readily available in nature, or they can be acquired from a sick person."[24] These qualities we have ascribed to microbes over time—life-giving, disease-bearing, mutating, evolving, and ubiquitous—now make it possible to assert that germs can also be instruments of terror.

GERM SOCIETY

Breaking the living world into a hierarchy of life has scientific and social consequences, as the management of that hierarchy is political. Recognizing that small life forms are vital to larger species does not mean they will be treated in any particular way; nor does it assume that all life will be equally successful at survival. The notion that one life must be sacrificed for the good of another plays out every minute on the African savannah and in every grocery store on the planet. One life form swallowing another might be regarded as not only a project to ensure the survival of a species but a process that ensures the continuation of life itself. Life is also productive. For example, humans labor in agriculture to cultivate plants, promoting the survival of multiple species. As a result, the production of life has been a politics of estrangement, creating a social system that values human life above others in order to convince humans that their labor is vital.

Marx posits a clear separation between human life and the other life forms that sustain it. These *other* lives might be called "nature," a resource that is essential to social life and one that sustains the human species:

> The universality of man appears in practice precisely in the universality which makes all nature his *inorganic* body—both inasmuch as nature is (1) his direct means of life, and (2) the material, the object, and the instrument of his life activity. Nature is man's *inorganic body*—nature, that is, in so far

> as it is not itself the human body. Man *lives* on nature—means that nature is his *body*.[25]

While Marx recognizes the productive work of humans in constituting nature through labor, he characterizes nature as something that is not alive, even though he considers inorganic nature an extension of the human body.

Though it relies heavily on a dualism between nature and culture, Marx's perspective reminds us that engaging with nature is a part of being alive, an amalgamation of the human body and a vast world of resources into a single working organism. The modern conception of a microbiome subsumes microbial natures into the human existence as a "direct means of life," for the work of microbes in digestion is part of the mechanism by which the human body extracts energy from nature after the labor in the field. However, the work of the microbiome requires new categorization, for microbes work in our guts for their own survival, with little interference from humans beyond the replenishing of food sources. As individual agents with their own intent, microbes resist being fully subsumed into this organic human body (placing them back in Marx's nature). Moreover, the assortment of microbes living in each human body is unique, constituted by an individual's interactions with varied environments. This demands a new categorization for microbes with their own agency and politics. The germs within us keep us alive, just as our bodies keep them alive. This symbiosis is inexplicable by either a collective or an individualistic view of the world, for each demands a privileging of either the whole or the parts. While our biological processes may seem individualistic—trillions of microbes doing the dirty work of digestion, only to be excreted from the body—the system can also be understood as a collectivity of germs and humans living and sustaining each other.

The nature of these interspecies interactions is further complicated by insights from genomic science. On the one hand, genomics has characterized individuality in terms of unique genetic codes. Human society or the holistic world might be understood to be the summation of uncountable numbers of genes and innumerable sequences of DNA. While slight variations in DNA make individuals, most genetic material is identical within a species. Thus, species can be understood by commonalities in DNA. Individual members of a species with genetic variations that make them weaker do not survive to pass on their genes: their elimination ensures the continuation and future fitness of the species.

Thus, genetics defines both individuals and species, binding them through commonalities and differences, but also threatens to rob individuals of agency in engaging with the world. Modern medicine, with its attention to personal fitness and individual disease prevention, provides a potential modification to the script of biological determinism. If people can overcome, or at least delay, genetic preselection through diet, disease prevention, and technology, they can pass on their genes to their offspring, while continuing to live an individual life that exceeds any genetically determined potential.

Microbes also play a prominent role in a narrative of human origin that casts nature as a selective breeder. Evolutionary theorists propose that complex life began when strong microbes engulfed other microbes and assumed their genetic traits. A microbe that could move might come in contact with a heat-tolerant germ, ingest it, and incorporate the prey's DNA into its own. Successive generations would be able both to move and to tolerate heat. By this theory, microbes could become ever more complex by contact with new microbes. By the theory of natural selection, populations that can resist certain stressors, such as disease carried by microbes, are able to survive to pass on their genes.

Suggesting that disease begets life seems paradoxical according to the individualist perspective, but if success is measured in the duration of the species, then the idea of disease weeding out the weak—a notion far more socially palatable when referring to nonhuman species like elk or trout—can be understood as a positive outcome. Scientists have described the net effect of billions of years of microbial attack as creating a diverse, complex, and vigorous human population. Culturally, this argument has been remade for political purposes through an alignment of disease with social behavior and morality. During the first wave of the AIDS epidemic, for example, the disease was frequently described as an expected and just consequence of thoughtless behavior or immoral acts, and the work of the human immunodeficiency virus was to cleanse the population morally. Different models, however, have the potential to open more equitable biosocial relations. Christopher Wills, for example, describes interactions with germs as an opportunity for species to continually change and grow.[26] A politics rooted in an understanding of microbes as creators of diversity might embrace broader cultural exchange between humans by embracing the positive outcomes of diversity in all forms.[27]

Locating germs within the human body has challenged historical formations of germs as external and foreign entities with associated meanings of malice and harm. More than a century ago, the germ theory of disease showed that contact with foreign pathogens could harm the healthy body. By proposing that disease did not originate within the body, the germ theory absolved humans from causing disease. Many prior conceptions of disease attributed illness to a moral failing or divine punishment for wrongdoing. However, microbes did not let people off the hook completely, for locating disease in the environment implied that it could be controlled, possibly through the same mechanisms that people were already using to labor and create their world. The development of sanitation as a cultural practice for managing the germ-ridden environment reworked the mechanisms for combating disease, drawing attention to the command of the environment.

This shift also elevated disease to the status of a social problem, in which disease emanated not from individuals but from certain groups, particularly those without the resources to create cleanliness. Contagion was materialized through interactions between humans. The germ theory arose at a time when increasing contact among human populations around the globe was changing the ways people encountered the world, and as David Raney says, "With germs (and germ theory) in the air, the fear of strangers took on an added dimension—their otherness could be catching."[28] The social consequences of the belief that contact with certain people meant exposure to disease has been shown by scholars to commingle formations of race, class, immigration, domesticity, and civic duty with the quest to evade the germ at the beginning of the twentieth century.[29]

As understanding of disease transmission grew, the separation of germs from the human body was blurred. Scientists showed that bodies could be carriers of disease without expressing symptoms. The germ had to be present to cause disease, but other individual and environmental factors were also implicated, undermining assertions that germs were inherently and indiscriminately bad. Responsibility was once again thrust on the individual, who was now responsible for personal fitness as well as environmental cleanliness. In the latter half of the twentieth century, Americans could find personal health advice through public health campaigns and corporate advertising, on the radio, in the supermarket, and from government representatives. Richard Lewontin argues that "problems of health and disease have been located within the individual so that the individual becomes a problem for society to cope with rather than society becoming a problem for the individual."[30] Because singular human behaviors assemble to create society, the unique positioning of the germ as both intimately personal and broadly social has led to the production of citizens who are simultaneously individually powerful and collectively vulnerable.

Microbes exist on a time scale and move in the landscape completely differently from previous formulations of nature. They represent a form of life that cannot be managed in the ways scientists once imagined it would be possible to control nature.[31] By altering our foundational understandings of what it means to be alive, the germ shatters seemingly evident categories of life and death. A virus, for example, does nothing unless it is integrated with a living cell. Because it does not metabolize, it would not be construed as alive by many theorizations of life. Yet if life is "mutation, reproduction and the reproduction of mutation," as famously defined by Hermann Muller, then viruses are alive: they perform these acts through the smooth reorganizations of their own biological matter to adapt and survive in endlessly changing environments.[32] By Carl Safina's definition, too, viruses are alive, for "in the biological arms race that is life on the Earth, success in reproduction is the only measure of success in living."[33]

The biologist Lynn Margulis has argued for a rethinking of life in the age of microbiology—not a new social understanding of nature, but new materialities that remake the world. She conceives of life as a "memory storing system" that bridges the physical with the historical, suggesting that the history of the world takes material form in biological life, a physical past that is producing the future.[34] This system creates a future that cannot be fully appropriated and subjects organisms in the present to a past they did not create.[35] Life is a "series of selves" that expend energy to exist but cannot be dissociated from past life or from each other. Thus microbes have rich social lives: they grow, reproduce, react to environmental insults, and form community structures.[36] This complex materiality opens the possibility of breaking apart categories of social difference by calling into question states of being as seemingly fixed as "alive" or "individual."

This persistent inseparability of human from microbial life, both materially and ideologically, is a guiding perspective for the chapters that follow. The forms of governance created to manage these worlds are the same, and they entangle nature in a political struggle that constitutes modern life. In the late nineteenth century Anton de Bary coined the term *symbiosis* to describe the situation of unlike organisms living together in relationships ranging from mutualistic to parasitic. The human connection to the microbe exists somewhere along this symbiotic continuum, not as a fixed point, but as an accumulation of the complex histories and practices storied in the bodies of humans and microbes alike. The management of the microbial world through the myriad practices of science, governance and citizenship has the potential to create futures that are as mutualistic and sustaining for human relationships as they are in the microbial world. However, the examples presented in this book caution that the relationships humans have with their germs are not generating hopeful futures. We will do well to face these microbial futures knowledgeably and sympathetically, because our relations with microbes are inescapable. As Margulis argues, "We can no more be cured of our viruses than we can be relieved of our brains' frontal lobes: we are our viruses."[37]

THE POLITICAL ECOLOGY OF GERMS

Jane Bennett advocates thinking of publics as "human-nonhuman collectives that are provoked into existence by a shared experience of harm."[38] Because microbes hold such potential for human harm (among many other qualities), they have led humans to rebuild their social systems. Thus microbes create new ecologies through their disturbance of the cultural order. The publics created by bioterrorism entwine people's political and biological futures with those of microbes. Such a future is neither inevitable nor determined but is constituted from natural histories that combine the cultural memory of disease with scientific practice and new forms of governance. I explore the contentious politics of bioterrorism to show that the work of science and government forges links between microbes and people in ways that create new identities, institutions, and inequalities.

This project examines how societies are governed when the gap between biology and the political lives of citizens grows indistinct. By producing organisms that are living, contagious, and unpredictable, the contemporary biosciences make it possible to conceptualize both microbial and human lives (and the hybrid form of an infected body) as security threats. Melinda Cooper argues that the "emergent possibilities of the life sciences" are of critical concern to the neoliberal project in the United States, with one outcome being the centralization of risk in governance.[39] New knowledge of nature creates new ideas about risk and new contexts for citizenship. Governments concerned with the care of citizens create science institutions that work to explain both threats and the efforts to contain them. In return for allocating public resources to science, governments require scientists to supply the state with knowledge about the public. Consequently, microbiologists are tasked with describing the effect of their work on social systems. This demands a particular type of work from scientists, whose tools remake landscapes in terms of the political context that sustains their work.[40]

War and science, two great pillars of modernism, remake each other through bioterrorism. The human body moves through a world of emergent and unspecified risks, and the continual production of knowledge of life generates new notions about how to harm the body.[41] A range of disciplines, from genomics and microbiology to public health and sociology, participate in the production of bioterrorism knowledge, bringing their own systems and institutions to the public contestations of risk. Science is a political practice, both the subject and object of power relations: the work of scientists cannot be separated from the systems they serve.[42] The desire for knowledge of human biological futures demands that scientists simultaneously produce both evidence of and solutions for bioterrorism. In the study of bioterrorism, scientists work in the realm of "anticipatory knowledge," projecting scientific calculations of the past events into the future. This type of knowledge offers governments and citizens bioterrorism scenarios that are scientifically ambiguous and tainted by cultural imaginings.[43]

Despite our intimate relations with germs—or perhaps because of them—our cultural interactions with microbes are persistently laced with fear. Fear is neither inevitable nor universal but an apparatus for shaping society, and thus intensely political.[44] As Joanna Bourke reminds us, fear is also emotional, and as such is experienced physiologically.[45] I do not claim that people's fears of microbes and bioterrorism are not real: rather I aim to draw attention to how those fears sustain systems of power. The particular ways that fears of germs have been embodied in bioterrorism brings those fears into the realm of human intervention, where they can be addressed through management practices. Through the production of fear, the security state justifies political acts that claim to minimize threats.

Though experienced in the present, fear is rooted in calculations of the past and imaginations of the future. In the fearful future created by bioterrorism, humans, microbes, and technology come together to inflict harm.[46] The event we imagine has no fixed endpoint, raising the possibility that our society cannot manage what it has created. Anthony Giddens argues that it is this preoccupation with the future, and particularly future safety, that generates the notion of risk.[47] The production of risk raises the question of who bears the "burden of survival" of biological threats.[48] When faced with vast, unspecified risks, people turn to larger institutions like governments to mitigate harm. Though risk is focused on the future, it demands action in the present, and because the future is not known, it enables governments to act even in the absence of a threat.[49] Andrew Lakoff uses the term *preparedness* to define behaviors that ascertain future threats and current interventions. He argues that when people generally agree that future risk demands action, they tolerate government interventions on their bodies.[50]

Foucault's theorizations recognize governance as both the acts that compel and coerce populations and those that create citizens who police their own behaviors in the interest of a shared future. Bioterrorism enables political responses that range from building new science labs and emergency centers to crafting television ads and video games. When the CDC issues a public service announcement during flu season featuring Sesame Street's Elmo telling us to "wash, wash, wash" our hands, the federal agency is engaged in work that both creates common recognition of threats and shifts the burden of preparedness to individuals. Other responses, such as stockpiling medical countermeasures for the entire U.S. population, fuse a state response with individual responsibility: the government will store and move vaccines, but individuals must make the choice to inject them into their own flesh. With other scholars of science and culture, I seek ways of understanding biopolitics on both the scale of unseen human-nonhuman intimacies and the broader scale of populations and disease.

Preparing for bioterrorism is not an inevitable activity in the genomic age, for we are continually remaking systems of power predicated on our understanding of life. Biological citizenship references the ways people use their bodies to claim rights and act politically to protect them. The contemporary politics of microbes has reconstituted citizens' claims to biological security. They turn to state institutions to protect their biological rights to health, safety, security, and life itself through a range of governing practices, such as regulating cross-border food systems or developing stronger and more effective vaccines. When the body wanders into a world of unspecified and mutable risks, it is awash in "unknown unknowns," and life becomes precarious. In this state, a new politics coheres around the body, which is striving for security amid emergent and changing threats.[51] The broad claim to biological security also enables governments to extend their authority through many social institutions, including the science and military complexes. This study of how bioterrorism remakes notions of biological citizenship contributes to a growing body of work that seeks to understand the multiplicities of citizenship created by changing ideas of nature and security.[52] The new microbiology creates new possibilities for biological citizenship by disrupting the fixity of life and nature. By delineating the material outcomes of bioterrorism in people's lives, I aim to give voice to biological subjects and show how lives are changed through bioterrorism, while identifying new commonalities in thinking about life, fear, risk, and nationhood in the modern age.

Every day modern subjects perform countless acts that constitute them as citizens while negotiating and legitimating equally wide-ranging government acts. In this historical and ethnographic search for what Adriana Petryna refers to as the "elements that unsettle and entangle people's lives," I seek to understand how new institutions emerge and find legitimacy within society.[53] The cultural forces of bioterrorism produce subjects who are as diverse and far-reaching as global networks allow, and the challenge of studying the modern biosecurity culture is to identify the connections that give bioterrorism form and tenacity in people's lives. Biological citizens in the twenty-first century live in a world where they can access a range of meanings and bring an array of experiences into the process of subject making. Their daily practices constitute the macrosystems of life, nature, and security.

Modern citizens are presented with a range of cultural fears and threats, but not all of them take hold in public politics. The story presented here moves throughout North America and covers a century of history to consider how bioterrorism finds form and stability within society. It illustrates how ideas of nature have worked across time and space to bring about consensus concerning biological threats to populations as well as practices that correlate governance with shared perceptions of nature and risk. This research retains ethnography's attention to daily practice while also examining the rhizomatic networks that complicate social relations in a global society. I have stepped back to look at the geopolitical scale and moved closer at points where meanings are stabilized and disrupted.

These shifts in time and space expose commonalities but also create broad collectives of human and nonhuman agents. Biosecurity practices have emerged precisely because modern landscapes are complex and publics are varied, and because the rapid production of knowledge of life and germs, when engulfed by global systems, reorders the world on scales both large and small. Kaushik Rajan argues that the ethnographer's traditional objective of "correcting" hegemonic and universalizing social theories by drawing attention to sites on the fringe falls short of describing precisely what those hegemonic forces are and the role they play in cultural formation.[54] This book attempts to account for the startling level of cohesiveness that the public exhibits in support of national biosecurity preparedness, as well as sites of contestation and cultural resistance, while attending to the individual acts that constitute both. The work of citizens involved in biosecurity—scientists, soldiers, presidents, emergency planners, doctors, activists, journalists—gives it prominence among the many messages of science, technology, health, risk, and fear that global citizens encounter. That the attempt to recognize the rapid saturation of knowledge in the modern world is an ambitious task does not lessen its import or the value of scholarly inquiry.

KEY TERMS AND A HISTORICAL GLIMPSE

Though public support for bioterrorism preparedness efforts is high, there is less consensus on the nature of those threats and the myriad ways to categorize the political acts that are part of biosecurity. As the idea of bioterrorism cohered toward the end of the twentieth century, an associated lexicon emerged, drawing terms from the life sciences and fusing them with political acts like war and terrorism. Deconstructing the use and evolution of a few terms exposes some of the cultural ideals about nature expressed in the language of bioterrorism. Defining and distinguishing among *bioterrorism, biological warfare,* and *biological crime* involves questions of biology, scale, intentionality, and context. All three categories of attack concern the use of biological life to create harm, though there are key differences in the actors and effects. Bioterrorism and biological warfare have large-scale effects, whereas biological crimes generally target individuals. *Biological warfare* is commonly understood as the use of biological weapons by a nation-state to cause loss of life and gain military advantage; *bioterrorism,* in contrast, aims to "disrupt our way of life and make us acutely aware of our vulnerability."[55] Intent matters, for biological harm can be caused unintentionally, for example through agricultural practices that introduce disease agents into a water supply. Definitions of bioterrorism vary among institutions because of the varying contexts of their work. For example, the CDC, which is primarily concerned with public health, has codified definitions of bioterrorism in terms of the disease-causing agents—the microbes themselves—rather than the character of biological weapons.[56]

Defining *biological weapons* and *weaponization* is complicated by the range of possible types and categories. It demands attention to the method of delivery, the target of the weapon, and the process of creation.

Anything from a sophisticated anthrax bomb to an intentional sneeze splattering germs on the subway might be called a biological weapon. Toxins, or poisons originating from plants, have been used as weapons, most famously in attacks using ricin, but definitions of biological weapons generally emphasize the use of pathogenic agents. Lajos Rozsa has proposed a definition of "biological weapons sensu lato" to include tools of aggression "[whose] acting principle [is] based on disciplines of biology . . . but excluding those based on inorganic agents. Synthetically produced equivalents (not necessarily exact copies) and mock weapons are also included." Rozsa builds his taxonomy of weapons types according to whether the causative agent in the weapon is created using theories and principles of biology: his definition takes account of the methods and ideas used to construct the weapon as well as the organic components. He further categorizes weapons based on two attributes: the size of the human population the weapon can harm (a military perspective) and the type of biological effect (a scientific perspective). Impact can be measured in terms of lives affected and duration of biological effect, and types are sorted by those intended to harm crops, those that affect human bodies but don't cause disease, and then by whether the weapons bear nonlethal or lethal diseases. I use the term *weaponization* to refer broadly to the act of making something into a weapon, though Rozsa illuminates the specific technical challenges of making a weapon, such as how to store and deliver pathogens to the intended target without harming "friendly forces."[57]

I use a range of terms in reference to microscopic life, most prominently *germ, microbe, pathogen, bacterium, virus,* and *microorganism.* The scientific literature shows variation in the precision scientists ascribe to and expect from these terms. In this book, I use *germ* and *pathogen* to reference a microorganism that causes harm to human life. In writing more generally about microbial life, I favor the term *microbe.* When appropriate, I use the category of *bacterium* or *virus* to delineate particular microorganisms.

Although this volume does not attempt to give a full accounting of the history of biological weapons use around the globe, I mention here a few key events for reference.[58] Early acts that might be considered forms of biological warfare include the use of infected corpses to spread diseases among enemies and the oft-told tale of Europeans' deliberately infecting Native Americans by sharing blankets used by smallpox sufferers. Such stories are foundational, suggesting that the inclination to use biological weapons long preceded our current armory. Some theorists have even argued that the intent to do harm is evolutionarily developed from early life forms that preyed on others to ensure their own survival.[59]

Several nations were engaged in offensive bioweapons research after World War I: not only the United States (see chapter 2) but also Great Britain and Japan. In 2002, Japan admitted that its army had released fleas infected with bubonic plague and tainted wells with cholera along China's eastern coast during World War II. Behind the Iron Curtain, it seems the Soviet Union continued to run an offensive weapons program long after it ratified the Biological Weapons Convention in 1972. The secretive weapons laboratory on Vozrozhdeniya Island in the Aral Sea has become the material of science fiction. Ken Alibek's desertion from the USSR as head of the bioweapons program, recounted in chapter 1, must be situated in the context of the global suspicion surrounding the facility's supposedly peaceful purposes—reported by Alibek to be fabricated to cover up the production of weapons. Questions about whether Iraq had a bioweapons program at the start of the first Iraq War, and further questions about whether Iraq used biological weapons against Iran, are characterized by similar suspicion.

Recorded acts of bioterrorism are rare. Aside from the anthrax attacks in the United States in 2001, two events seem to fit this category of biological attack. In 1984, the Rajneeshee group in Oregon spread strains of salmonella on salad bars in order to influence a local election. The group hoped that stomachaches would confine voters to bed on Election Day, allowing the Rajneeshees to sway the numbers in favor of their candidates. They sickened about 750 people. In 1995, a cult in Japan, Aum Shinrikyo, released sarin gas on the Tokyo subway, killing twelve people. This incident is often regarded as an act of bioterrorism, though sarin is a chemically derived nerve agent. This relatively short history of documented biological attack seems disproportionate to the cultural effects of bioterrorism I describe, giving further credence to the claim that bioterrorism creates potent social effects precisely *because* the direct effects of bioterrorism, in the form of microbes intentionally infecting human bodies, are unknown.

OVERVIEW OF THE BOOK

The following chapters explore what is at stake in the production of bioterrorism and the ongoing work to prepare for biological attack. I examine places across North America where bioterrorism has arisen as a dominant narrative and where associated meanings have been stabilized and contested. Some sites seem obvious, such as the CDC, but controversy also surfaces at unexpected sites, like a high-tech research facility in a remote Montana town. Any one of these research sites could be the subject of an in-depth ethnography. I do not assert a thorough knowledge of all the security practices of the CDC or the Department of Homeland Security. Rather, by surveying a range of sites, I hope to point out certain practices and beliefs that are common to all of them, giving support to larger claims about the social effects of bioterrorism and demonstrating how political ecology can make sense of the ways nonhuman species disrupt our social systems.

Germ Wars begins with a close look at microbes. Unseen organisms brought animated nature into all human environments, through processes and phenomena we describe using terms such as *contagion, mutation, adaptation, species,* and *life.* The origins of the germ theory of disease in the nineteenth century, which produced new human and nonhuman collectives, changed how people viewed nature. Scientists looked closely at microbes to better understand their behaviors. This knowledge produced not only new fears but also the promise that these organisms could be managed, contained, and even eliminated for the preservation of human life. The hope that disease could be eradicated from the globe laid the foundation for a new politics of public health.

Chapter 1 examines how the technology of vaccination was used to eradicate smallpox. A "war" against smallpox was waged internationally, infusing the work of public health with the discourses of conflict and a global politics of race, class, and gender. Ultimately, containing the virus required controlling humans. The campaign thus demonstrates how governments can access and alter individual bodies in the name of public health. Further, public health is ripe with metaphors of war, inscribing the human relationship with microbes in terms of conflict and opposition. This battle plays out in everything from domestic sanitation to the modern infatuation with disinfectants and instant hand sanitizer. Even though smallpox no longer occurs naturally on the planet, the smallpox disease narrative is still invoked to sustain political fears of terrorism and justify national security actions. Chapter 1 concludes with a consideration of biosecurity institutions emerging from the Cold War that now perpetuate the endless war against disease, including the smallpox virus that no longer exists in nature.

Waging war *against* germs simultaneously created the possibility of waging war *with* germs, creating industries of biowarfare alongside public-health programs. As the U.S. military establishment took up the study of the most dangerous pathogens following World War II, an alliance between microbiology and warfare was created in top-secret, high-security institutions. In chapters 2 and 3, I examine the practices and spaces created by the allied industries of war and science that materialize the belief that nature can be contained.

Near the end of World War II, the United States responded to reports that enemies abroad were building weapons from microbes by establishing a facility for bioweapons research at Fort Detrick, Maryland. The scientists at Fort Detrick had to create technologies that would contain lethal pathogens while keeping them alive for study. Like vaccines, these new laboratory spaces materialized a promise that microbes could be managed by humans. Scientists also worked to expand the possibilities of microbes to destroy human life, broadening popular imaginings of microbes as killer germs.

By the time the United States terminated its biological weapons program in the 1970s, the Department of Defense had expanded public knowledge and imaginings of what microbes could do to harm populations. I show how the U.S. desire to produce biological weapons led to the creation of high-security containment laboratories where scientists could safely study deadly microbes and to the idea of biosecurity as a set of technological and spatial practices that contain microbes for scientific study. Over time, biosecurity came to be understood more broadly as the containment of biological threats, a transference that emerged from and sustains a broader social belief that technology can contain and control natures.

The principles of biosecurity established at Fort Detrick came under public scrutiny decades later when citizens in Hamilton, Montana, protested the first high-security biological laboratory built after 9/11. In chapter 3 I describe the deeply divisive public debate that ensued, during which residents in a rural town demanded that the government secure the community against the risks presented by the deadly microbes under study. Looking at the hundred-year history of the laboratory in Hamilton reveals repeated discord between scientists' faith in technology and citizens' fears. Working through civic and legal systems, citizens of Hamilton formalized a cultural belief that the government bears responsibility for protecting people from the risks of deadly microbes.

The promise of science that germs can be controlled has rationalized government action and established the care of citizens' health, through such means as sanitation, vaccination, and education, as a responsibility of the state. Chapter 4 focuses on the work of bioterrorism preparedness, where governments manage diseases that are not present within the population, anticipating future events and then planning and rehearsing a response. In this effort the U.S. government has recruited public health institutions to the work of homeland security and defense. Political transformations at the CDC show that mechanisms of health surveillance and incident response have been remade for the bioterrorism crisis, changing the role of government in disease management overall as well as the mandates of the nation's largest public health agency.

Because no large-scale bioterrorist attack has taken place in the modern age, planning for such events requires imagination. Simulations, ranging from computer-generated models to role-playing events involving thousands of human actors, are used to create information about biological attacks. All these efforts rely for their credibility on scientific information about microbes. In chapter 5, I argue that this work of simulating bioterrorist events creates new systems of scientific knowledge and new ways of knowing nature. When simulation is accepted by a community of planners and policy makers as a scientific means of predicting both human and microbial behavior, it changes the calculation of risk and the types of actions taken to deter imagined future events. As such simulated future natures gain credibility, it seems germane to ask how modeling has been adapted as scientific practice and how it has gained tenacity in the current political climate. I particularly question the use of scientific predictions of microbial behavior to rationalize political actions that manage human life, individually and collectively.

The final chapter describes how the boundaries of the nation-state are asserted through the imagination of microbial nature. The goal of national biosecurity is to restrict movement of pathogens while still enabling the flows of goods, services, and human and nonhuman life. Microbes cling to other forms of life, whether they are travelers, imported food sources, or migratory birds. Defending borders against biological threats defines the nation in terms of how living natures move and interact with geopolitical environments. I conclude with a case study of border security in the U.S. Southwest, showing the complex political natures created by the policing of human and nonhuman bodies across the U.S.-Mexico border. The biopolitical policing of the modern microbe creates new ways for governments to care for citizens. Battles for justice are waged every day as people seek access to that care.

Bioterrorism is constituted today through the blending of specific histories of disease and violence with modern biological science. Regardless of whether a bioterrorism threat has ever existed, the fear that microbes will harm humans is remaking social systems and altering modern life through the work of preparing the population and securing our bodies against biological threats. Incorporating new conceptions of life, biology, and risk into our ways of knowing their world and living in it has the potential to open possibilities for more equitable futures. Recognizing bioterrorism as a produced event that performs a particular function within a modern political system shows the ways in which society has been organized for the production of violence in the name of biosecurity. Life is at the core of governance, and the complex and fluid materialities of life created by microbial natures contain the potential for rich and nuanced conceptions of human subjects.

Nature is at the center of public concern as never before, but the forms of nature are no longer recognizable in a traditional sense. One of the central sites for the production of nature over the last century has been the massive expenditures on disease control. Political ecologies must take seriously the work of managing microbes as an environmental practice sited in laboratories, health centers, and government offices. Studying these management practices illustrates the complexity of modern environments and natures. In particular, it informs debates about the nonhuman and the cultural production of fear through science practice and governance. Including microbes more fully in the critical study of nature will further elucidate how institutions of health, war, and science have been built around ideas of nature, remaking the possibilities for biological citizens and environmental futures in the twenty-first century.

1

"Smallpox Is Dead"

The Public Health Campaign to (Almost) Eradicate a Species

> Disease has long been the deadliest enemy of mankind. Infectious diseases make no distinctions among people and recognize no borders. We have fought the causes and consequences of disease throughout history and must continue to do so with every available means. All civilized nations reject as intolerable the use of disease and biological weapons as instruments of war and terror.
>
> —President George W. Bush, "*Strengthening the International Regime against Biological Weapons*," November 2001

In February 2002, President George W. Bush announced that the United States government would spend $6 billion on bioterrorism defense. This, the largest single bioterrorism expenditure in U.S. history, was followed by expanding allocations of resources over the next decade. In the wake of the anthrax scares in September and October 2001, Bush had declared disease to be the "the deadliest enemy of mankind."[1] The president situated the anthrax attacks, and bioterrorism more generally, in a cultural history of disease that construed the microbe as a natural enemy of human life, thus bringing value systems centered on health and care into the service of a national biosecurity state. Bush naturalized acts of war through a narrative of human survival against disease.

Though they seem to have emerged separately, the rise of the modern security state is entwined with the rise of a new genomic biology. National security operates through the continual production of risk, and microbiology provides a potent site for cultivating risk by creating a pervasive nonhuman nature with a proclivity of its own. The dynamic and volatile qualities of nature, reinvigorated in the new millennium by biotechnologies and knowledge of microscopic life, provide a potent site for rethinking what citizens require of government in a world of unspecified risk.

Centuries ago, when scientists began looking at the world through microscopes, they began to shape the nature of the microbe according to the fears and desires of human societies. Social fears have become part of the microbe's material form and existence. Scientific study and technological innovation also inscribed in microbes the promise that they could be managed, contained, or even eliminated for the betterment of human society.

The management of disease on both the national and the personal scale has long been a component of the relationship between states and citizens. However, the promulgation of the germ theory of disease from the mid-seventeenth to the late nineteenth century specifically located the origins of disease in nonhuman life forms moving unseen around and within human bodies, changing how disease might be governed. The careful management of environments promised to separate humans from germs and even to eradicate species to alleviate human suffering. The scientific creation of microbial agents of disease fashioned a nonhuman enemy with its own volition. Because the microbes were alive, because they acted according to their own purposes, and because their invisibility and pathogenicity imbued them with an element of trickery and deception, the vibrant germs became a model enemy for militant action. The war on disease created a state of exception in which human rights could be subordinated to the overarching objective of eliminating infectious disease. This history underpins the twenty-first-century declaration of a war on (bio)terror.

When he announced the bioterrorism preparedness expenditures, Bush claimed, "History has called us into action."[2] In targeting the nonspecific threats of terror and biology, national security practices invoke the historical formation of disease. Because there had been no major bioterrorism events in the United States except the anthrax attacks four months earlier, the president's speech reworked deeper histories of disease and military might. Bush described military successes using disease surveillance during the Korean War, proposing to "adapt" these technologies to a new mission of domestic surveillance. He cited disease-eradication programs and promised that bioterrorism preparedness would create "some incredible cures to diseases that many years ago [people] never thought would be cured."[3] Conflating war with public health normalizes a state of emergency, disguising as quotidian practice the violations of privacy and liberty that might arise from state surveillance.

Contemporary ideas of national biosecurity are built around the modern microbe, a biological agent created over time through the work of scientists and consumers, public health institutions and governments. The presence of microbes within the human body, the collective experience of vulnerability, the communal qualities of contagion, and the hybridity of techno-natures created by the mechanical alteration of microbes form a compelling backstory for bioterror. Disease-control practices created a story of nationalism that imbued the microbe with global politics. These histories of biology matter because they are at the forefront of the many narratives available to citizens in assessing future threats. Humans draw from a range of personal and collective experiences to assess their own vulnerability and make sense of biological risk. Sheila Jasanoff defines the human effort to calculate risk as "our paradoxical attempts to cope with the irrational in rational terms." She argues that people extrapolate from their experiences of the past to predict misfortunes in the future. More than simply imaginations of the future, "risk is a disciplined projection of archived historical memory onto the blank screen of the future."[4]

In this chapter, I present a cultural history of one virus, *Variola major* or smallpox, to show part of this archived memory of disease that modern Americans project onto their biological futures. I peel back the layers of experiments, technologies, field work, legislation, and values that have cohered around this microorganism to show how the smallpox eradication program viewed containment as the ultimate form of human-germ politics. The program enacted a belief that humans can manage nonhuman life and established an authoritative role for governments in shaping microbial nature. It also demonstrates how humans reshaped the ten-thousand-year-old *Variola* virus, dramatically altering its global presence and then remaking the virus in laboratories and by synthesizing DNA sequences. The biopolitics of smallpox today exemplify how people imagine the future of microbes based on social histories of disease and science.

In the absence of a major historical bioterror event, people seek other rationalizations for their fears of bioterrorism. An organism like smallpox carries meanings of vulnerability and contagion while still being open to new meanings. Precisely because the virus is both contained in laboratory freezers and continually remade in cultural discourse, smallpox has become a potent actor in the biopolitics of terrorism. While the world was engaged in a global war against smallpox in the middle of the twentieth century, the United States was also rehearsing the logics of containment in the Cold War. Through this conflict, Americans imagined a totalizing threat and experienced the militarization of civilian daily life. The chapter concludes by examining how the establishment of the U.S. Department of Homeland Security (DHS), as an institution of the modern security state, created new biopolitical subjects for the war on terror. These narratives illustrate how government systems take shape around fear, laying the groundwork for further conversations about the natures of national security. Bioterrorism, rooted in the ongoing politics of life, risk, and government, binds natural histories to the politics of war. This chapter lays the foundation for exploring the transformations wrought by war and science on the socioecological relations between humans and their microbes. The work of managing threats has created material natures that change the calculation of risk for citizens. This story begins with a cultural history of execrable disease, singular immunity, and collective social action.

THE DEADLIEST VIRUS IN HISTORY

> Smallpox was always present, filling the churchyard with corpses, tormenting with constant fear all whom it had not yet stricken, leaving on those whose lives it spared the hideous traces of its power, turning the babe into a changeling at which the mother shuddered, and making the eyes and cheeks of the betrothed maiden objects of horror to the lover.
>
> —Lord Thomas Babington Macaulay, *The History of England from the Accession of James II*, 1800

Though the virus would not be seen under a microscope for another century, Macaulay's vivid description of the presence of smallpox in English society indicates the grim fear created by the disease. Smallpox has been responsible for much of the suffering, blindness, scarring, and death in human history. The disease is only moderately contagious, requiring close contact with an infected body to spread to a new host, but it is fatal to 30 percent of people who contract it. Historians estimate that the smallpox virus killed three hundred to five hundred million people in the twentieth century alone.[5]

Variola major has long been known by its expression on the body, and much of its cultural effect comes from its gruesome transformation of human flesh. Flu-like symptoms emerge within two weeks of infection, and then small lesions appear in the mouth, growing and rupturing, spewing the virus into the body through the saliva. After this surge of infection, the characteristic rash emerges, with pustules—smaller than those of syphilis—developing during a final phase, descriptively named "ordinary," "flat," "modified," or "hemorrhagic." Lifelong scars mark bodies that have hosted *Variola major* and survived.

Jessica Stern describes bioterror as a "dreaded risk" because its depiction of the future hinges on experiences of infection that evoke visceral horror and cannot be avoided or expelled.[6] Smallpox produces just such a reaction. However, the worldwide campaign to eradicate smallpox and the integration of smallpox prevention into the public health economy have produced another way of knowing smallpox, in terms of social control and the management of collective life. As modern science practice revives smallpox with narratives of a transgenic mutation of the virus that could escape control, the primal experiences of disgust rise again.

Smallpox's origin story narrates both the benefits and the consequences of human intimacy with nonhuman nature. Domestication brought livestock into human societies, putting people in physical contact with mammals and birds and their accompanying microbes. Crowding among humans and animals then enabled the disease to spread.[7] This narrative of infectious disease emerging as a consequence of living too closely with animals persists today in discourses about avian flu in Asia or swine flu in Mexico. Biologists and historians theorize that continued interspecies interactions over time build a population's collective immunity, making individual bodies less susceptible to disease and inscribing in them a genetic code that is transmitted to successive generations for the survival of the population. Domestication of animals made the disease, but the interspecies encounter also remade the human body such that it could endure infection.

This genetic rewriting of humanity is often used to describe the European body on the verge of colonial expansion: teeming with germs but empowered by its own immune system. The vulnerability of native people has also been written into the landscape. Stereotypes of nomadic, game-following bands of people moving through wild lands without domestic animals depict the opposite conditions as Europe. Under these circumstances, native people could not cultivate immunity and became vulnerable to foreign diseases. This simplistic, deterministic argument exemplifies how readily the biological experience of nature is used to explain imperialism as a natural process and thereby naturalize unjust social relations.[8] That scholars and citizens alike have used microbes to explain the outcome of the colonial encounter indicates the power we ascribe to germs to shape relationships.

The smallpox virus has been credited with powerful transformations of human society through narratives that naturalize cultural injustice. Historians attribute the Spanish conquest of the Aztec empire in the 1520s to the disease, because the more-numerous Tenochtitlán warriors contracted smallpox from Spanish corpses and succumbed to the disease.[9] From 1520 into the 1800s, smallpox spread throughout North and South America, Australia, and the Pacific, leaving behind diminished empires—Aztec, Inca, Cherokee, Eskimo, and others—where peoples devastated by disease could not defend their homelands. Scholars estimate that measles, flu, and smallpox killed up to 95 percent of the native populations.

The lack of immunity to European disease, an outcome of a particular relationship with nature that did not expose them to animal diseases and build genetic immunity over time, is characterized in these narratives as weakness. Describing the colonial conquest in these terms displaces the violent act onto the microbe and attributes a racial superiority to a particular genetics produced by living at close quarters with animals. In this accounting of history, microbes remake human genetics to create cultural power. Although the genetic perspective came centuries after the colonial encounter of the Americas, the social effect of disease was known at the time. Some saw it as an unfortunate consequence of cohabitation, but to others it embodied divine will. As one Methodist minister declared, "Providence designed the extermination of the Indians and that it would be a good thing to introduce the smallpox among them!"—an opinion assumed to be shared by "most white people living in the interior of the country."[10] Vulnerability to disease was equated with weakness and the devaluing of human life.

Perhaps the overdetermined fear of bioterrorism in the United States can be partly explained by the belief that the nation itself was constituted through the spread of disease.[11] William H. McNeill and Alfred W. Crosby describe the devastation that smallpox brought to native people in the Americas as it spread ahead of the colonizing powers. Crosby estimates that native populations in what is now the United States were once comparable in size to the populations of Aztecs and Incans at the time of European contact. Though rich in resources, the fertile areas of the Southeast were comparatively deserted when white settlers arrived from France and Virginia. With no evidence of major ecological shifts, it seems likely that disease was the cause of this depopulation.[12] Because smallpox lingers in the body for up to two weeks before causing symptoms, the virus could have been carried along trade routes or between settlements in asymptomatic bodies, both European and Native American. Smallpox created the conditions for conquest. Current inhabitants of North America need not look far to see that disease can play a powerful role in changing regimes and shifting the undercurrents of war.

From this naturalized dynamic of racial warfare also emerge accounts of smallpox being manipulated as a weapon of war. The legend of British soldiers passing smallpox-contaminated blankets to native tribes in the Ohio Valley during the French and Indian War is often cited as an example of the early use of biological weapons. Though historians question whether a deliberate strategy of biological infection was ever used successfully (because smallpox was already known to be sweeping through native populations), the British may have attempted it. In 1763, when the Delawares laid siege to Fort Pitt for more than a month, the British general Jeffrey Amherst suggested, "Could it not be contrived to send the small pox among the disaffected tribes of Indians? We must on this occasion use every stratagem in our power to reduce them."[13] Colonel Henry Bouquet offered to "try to inoculate the bastards with some blankets that may fall into their hands, and take care not to get the disease myself."[14] Notably, the flight of people from nearby communities to Fort Pitt for refuge had led to cramped quarters that facilitated the spread of disease. The officer in command of the fort wrote, "We are so crowded in the fort that I fear disease . . . ; the smallpox is among us."[15] The soldiers sought a measure of justice by transferring the disease that was crippling them to their enemies.

While taking advantage of a truce to pass an infected blanket to an enemy seems a reprehensible and inhumane act, the frequent retelling of this incident in bioterrorism history indicates that the circumstances resonate with contemporary notions of biowarfare. A group under siege saw a weakness in their enemy and exploited it. When policy makers describe the United States as vulnerable to bioterrorist attack, they, too, assume that terrorists in oppressive situations with no other weapons will see vulnerability in the nation's inevitable susceptibility to disease and use a microbial weapon to inflict harm. The narrative of the smallpox blankets circulates in the cultural history of bioterrorism with two additional effects. First, it associates bioterrorism with the colonizing of America and the succession of native tribes by European governments, providing longevity to the bioterror crisis while entrenching it in notions of nationalist identity. Second, it suggests that biological warfare does not require privileged science or technology to construct systems of delivery.

NATURE'S VACCINE

The worldwide spread of smallpox exemplified the different possible effects of interactions between microbes and bodies. If the conditions that provided immunity to certain populations could be deliberately created and distributed, then vulnerability and resistance might come under social control. Inventing mechanisms to create immunity could shift the dynamics of disease, bringing vulnerability into the political calculations of governments. The cultural history of vaccination shows the politics of wealth and nature that enter into health systems through the technological production of immunity.

Long before microbes were identified as vectors of disease, societies organized to protect their citizens from pathogens, using mechanisms ranging from home remedies to militant quarantines. In Asia, people inhaled or swallowed powdered smallpox scabs to produce a degree of immunity from smallpox. A practice called variolation spread to Europe from Turkey, where people achieved a level of individual immunity by rubbing the liquid from a smallpox pustule over a scratch made on the arm with a needle.[16] Milkmaids and farmers who worked closely with their cattle claimed immunity to smallpox; this bit of folklore spurred an apprentice physician, Edward Jenner, to develop the first truly effective technology to prevent infection with smallpox. Jenner injected his gardener's eight-year-old son with liquid taken from a cowpox blister on a milkmaid's hand and a few weeks later variolated him with smallpox. The boy developed a cowpox lesion but never contracted smallpox. Jenner continued his research with cowpox, injecting children throughout his neighborhood and eventually publishing his limited but conclusive evidence in a self-financed pamphlet.[17]

That Jenner was allowed, and even encouraged, to experiment on local children, intentionally exposing them to a deadly virus, testifies to the blight that smallpox was on the population. If children were likely to contract smallpox anyway, why not expose them while they were healthy and stood a better chance of survival? Jenner named the disease showing up on dairy farms *Variolae vaccinae,* meaning pox of the cow, and called his inoculation process *vaccination.*[18] Though Jenner's work was received with skepticism by his peers and has more recently been criticized as unscientific and even unnecessary, within ten years the technique had spread throughout Europe, Asia, and the Americas. Because of its lower fatality rate, vaccination quickly replaced variolation as the preferred form of immunization against smallpox.[19]

GENERAL VACCINATION DAY AT THE PARIS ACADEMY OF MEDICINE.

FIGURE 2. An April 23, 1870, article in *Harper's Weekly* described the fashionable "General Vaccination Day at the Paris Academy of Medicine": "Thousands of arms are presented to the doctor's lance[, and] utmost care is exercised in obtaining the vaccine matter from a heifer selected by the medical authorities."

Within three years of Jenner's publication, one hundred thousand people had been vaccinated in Britain, but problems with distribution of the vaccine surfaced quickly.[20] The smallpox vaccine is rare in its use of a live virus. In Jenner's day, the vaccine had to be extracted from an active rash and had limited viability in storage. For the first half of the nineteenth century, the vaccine was passed from the scab of one vaccinated individual to the next, occasionally boosted by an injection of fresh pox straight from the cow. Around 1840, a technique for producing large amounts of vaccine in cows became popular, to the extent that doctors brought infected calves into their offices and scraped the live virus right off the animals' flanks. This newly technologized relationship between humans and animals again changed the nature and sociality of immunity. A woodcut illustration published in *Harper's Weekly,* "General Vaccination Day at the Paris Academy of Medicine," depicts the process as a public spectacle. Men in top hats look on while women bare their arms for the physician's needle. Strapped to a table, amid the throng of people, lies a cow, the source of the vaccine matter. Notably, the accompanying article, written for a U.S. audience, uses the festive appearance of the gathering to argue for the safety and popularity of vaccination, which was still viewed skeptically in the United States.

Human bodies as well as animal bodies enabled the global spread of the smallpox vaccine. After receiving the vaccine in Geneva in 1800, King Charles IV of Spain sent his physician, Francis Xavier de Balmis, on an expedition to take the vaccine to Spanish America. Balmis brought twenty-two orphans on the voyage, vaccinating two boys every ten days to keep the vaccine alive in their bodies.[21] Once serving solely as carriers of disease, human and animal bodies now became vectors for the cure by transporting live viruses in living flesh. Eventually, the smallpox vaccine would be separated from bodies, distancing vaccination from its origins in nature and transforming it into a commodity for trade and a tool for the security state.

In the early 1800s, attempts to sell the smallpox vaccine for profit in the United States proved unviable because it could be harvested from anyone expressing the characteristic cowpox lesion. All a physician needed to obtain a supply of the vaccine was a body recently vaccinated or an infected cow, along with the knowledge of how to transfer it between hosts. The government became involved in the vaccination project in 1813, when James Smith was appointed federal vaccine agent and charged with "the preservation of the genuine vaccine matter for the use of others."[22] The government thus began regulating the authenticity of the vaccine, creating a scale by which value could be assigned and traced.

Within a decade, an incident involving the mislabeling of smallpox virus as vaccine, which resulted in the death of ten individuals, brought the government's role in vaccine distribution under scrutiny. Representative Hutchins G. Burton of North Carolina began a campaign to repeal the act that had established the agency, calling it "a mere nuisance, of the most dangerous kind, . . . suffering, under the authority of our laws, hundreds to be slaughtered with indifference."[23] Burton attacked the agency as a monopoly where the government was enabling one man "to accumulate wealth, by levying contributions from all parts of the union."[24] When a smallpox epidemic struck Baltimore, Smith's hometown, a local periodical claimed, "The act of congress to encourage vaccination, has rather, in our opinion, tended to encourage small pox, by making a matter of individual profit out of what had better been left to the general care of medical gentlemen," and later, "There is something wrong or rotten in this business."[25] The controversy eventually brought about the dissolution of a federal authority over vaccination, effectively returning the responsibility for vaccination to medical professionals and individual citizens. More than a century would pass before government again asserted authority over the mass vaccination of citizens through programs aimed at eradicating smallpox worldwide.

Though medical professionals retained purview over the production of immunity, the practice was deeply political, requiring the reconfiguration of social values. Ed Cohen's study of immunity explores the controversy surrounding inoculation in colonial America, concluding that it brought overdetermined values of nature, God, and governance into collision. An immunized body is a hybrid of natural and social laws and praxis. Cohen observes that immunity practices are "hybrid networks," binding the human and nonhuman in such a way as to render their connection invisible.[26] Indeed, as physicians stopped scraping vital matter off of living bodies to inoculate patients, the natural sources of immunity became disguised, allowing medical protocols like vaccination to assume a cultural position in opposition to the diseases of nature.

Despite the natural history of vaccination, immunity constitutes and depends on a belief that our bodies separate us from other humans and the world around us, rather than binding us to our environments and each other. The immunized body requires new scientific explanations of the self in which it is constituted not as contiguous with the environment but as an assemblage of many parts that can be acted on independently with great effect upon the whole. Cohen observes that "after the advent of immunity-as-defense, bioscience affirms that living entails a ceaseless problem of boundary maintenance."[27] The conviction that our bodies are separate and distinct from the world around us allows us to locate threats to human life in the world on which we materially depend for our existence. Biological threats draw power from this distinction, which also shapes the intervention of political systems to police bodies to ensure the survival of populations.

ERADICATION

> But, now when . . . the human frame, when once it has felt the influence of the genuine cow-pox in the way that has been described, is never afterwards at any period of its existence assailable by the smallpox, may I not with perfect confidence congratulate my country and society at large on their beholding, in the mild form of the cow-pox, an antidote that is capable of extirpating from the earth a disease which is every hour devouring its victims; a disease that has ever been considered as the severest scourge of the human race!
>
> —Edward Jenner, *An Inquiry into the Causes and Effects of the* Variolae Vaccinae, 1798

In the closing lines of his vaccination proclamation, Jenner predicted that vaccination might be used to eradicate smallpox. As he foresaw, application of the vaccine rapidly reduced the incidence of smallpox across the globe, and mandatory vaccination programs eliminated the disease in most countries by the mid-twentieth century. However, millions of people still carried smallpox, and in 1967 the World Health Organization (WHO) began an intensified smallpox eradication program with the aim of eliminating the disease. The WHO campaign has been lauded as a lesson in global disease control and an example of innovative thinking, cooperation, and goodwill to alleviate worldwide suffering without regard to borders or politics. Disparagers claim that smallpox was already on its way out and that the campaign capitalized on the waning of the disease to expand the reach of public health.

Regardless of its motivations, the WHO campaign mobilized people against smallpox, taking up technoscience weapons to eradicate the disease. The campaign constructed smallpox as an enemy to be fought, controlled, and eliminated, and it produced the weapons, strategies, tools and attitudes that continue to be deployed in modern public health campaigns and the logics of biosecurity. Furthermore, the eradication of smallpox from nature was political work that created enduring new formulations of difference with respect to vulnerable and risky populations.

At its annual meeting in 1958, the World Health Assembly, WHO's governing body, adopted a resolution calling for mass vaccination to dramatically reduce the incidence of smallpox in the population.[28] The plan to vaccinate millions of people in countries where smallpox was endemic held much political appeal: many countries had infrastructures in place for vaccinating, administering vaccinations would provide an economic boost, and it was a clear show of government engagement to improve public health. Although smallpox was eradicated in countries with means, interest in the program waned when mass vaccination proved to be expensive and time consuming. Programs instituted by the United Nations over the next decade were unsuccessful until World Health Assembly Resolution 20.15 laid out a new economic rationale for smallpox eradication. Pointing to the billions of dollars spent in developed countries to keep their populations immune and their borders secure, the resolution endorsed global eradication as a money-saving option. The resolution passed.[29] Andrew Lakoff points out that the antismallpox campaign aligned with Cold War ideals intertwining public health with economic progress.[30] The investment of nations where smallpox was already eradicated underscored the idea that sustaining citizens' right to health was a work of global security.

The eradication plan depended on the willingness of all nation-states to elevate the status of developing nations through the improvement of health systems. To succeed, the campaign needed access to the bodies of millions of people in more than forty countries, many of whom were also tormented by poverty, civil war, and a range of other health concerns. Even as it targeted individual bodies and microbes, smallpox eradication performed a politics of race and class on a global stage.

The language of the campaign borrows generously from the dictionary of war, describing the public health task as a military invasion and producing a corpus of combat terminology aligned with disease. The program's director, D. A. Henderson, emphasized that "WHO had no authority, other than that of moral suasion, to compel any country" to participate, but it saw public health as moral imperative that transcended any political considerations.[31] Those enlisted to eliminate the disease spoke as if they were going to war, though Henderson claimed that the limited finances and lack of military authority differentiated the undertaking from the popular perception of a war campaign.[32] As mass vaccination proved infeasible, a new strategy was evolving, one to target and break the chain of transmission. One WHO official recalled,

> It was on a hot, blistering June afternoon in 1973 that the "war plan" that eventually spelt victory over smallpox in India was set in motion. Till then, the relentless war against an enemy that knew no mercy had not been going on too well. If anything, it had become a general's nightmare. Though there was no dearth of "troops" or "ammunition," the problem was to get them to the right place at the right time. Naturally, the casualties were heavy—over 16,000 reported dead and more than five times this number maimed and disabled.[33]

The new "war plan" involved an approach in which a "ring" of resistance was created around an infected locale by quarantining and vaccinating residents of the surrounding areas. In the words of Rabindra Nath Basu, the Indian National Smallpox Eradication Programme officer, "We decided then that instead of expending our resources against the entire enemy forces simultaneously, we would concentrate on their strongholds."[34] The restructured program recruited "officers" and "advance teams" who were put through "highly intensified training courses to qualify . . . as experts" in detecting smallpox.[35] These teams conducted "reconnaissance trips" to identify "enemy" areas.[36] When an outbreak was reported, the team would "blitz" the area with "vaccination devices and vaccine—the guns

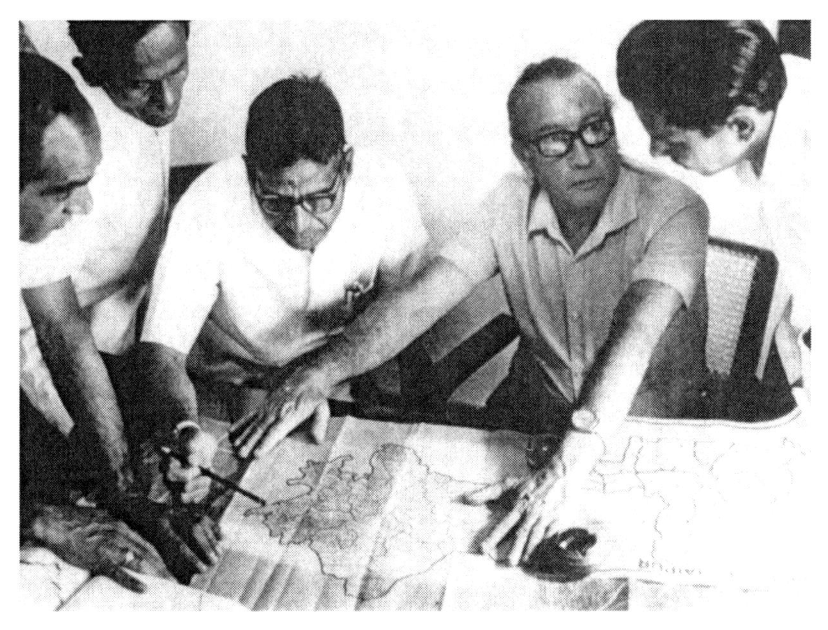

FIGURE 3. "Planning Strategy." Public health workers in Rajasthan plan their next move in the WHO campaign to eradicate smallpox. Image courtesy CDC Public Health Image Library.

and bullets of the campaign."[37] The campaigners had been immunized against smallpox, so while the virus could "wage war" on other people's bodies, it could not infect the public health "soldiers." The ring plan created a reactive response to an outbreak, a contrast to the preventive approach of mass vaccination, further instilling the urgency of combat into the overmilitarized campaign.

The war against smallpox was waged for more than a decade, creating an enduring state of exception in which acts of surveillance, secrecy, and civic duty shaped daily life. Surveillance of populations to detect infection was justified as vital to stopping the spread of smallpox. Containment also required that populations be trained to report the disease to authorities. Citizens had to be persuaded to act on a moral imperative, often violating relationships of trust and privacy associated with the sick and dying on behalf of some "greater good." Jitendra Tuli reports going into classrooms to explain the campaign to children, then asking them to report any diseases in their own homes to their teachers.[38] In later years, officials offered rewards to individuals reporting cases of smallpox. The monetary incentive turned up hundreds of false

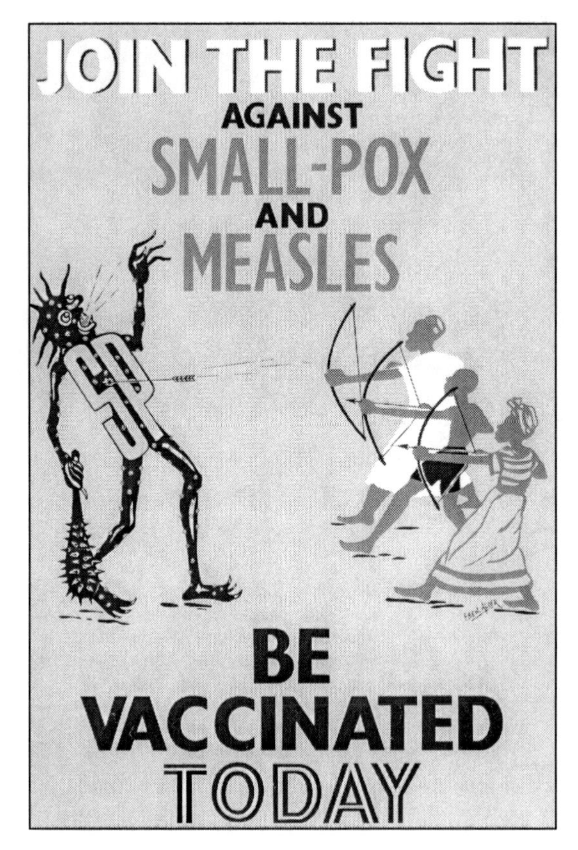

FIGURE 4. A smallpox vaccination poster from West Africa depicts a black-skinned, pockmarked, larger-than-human smallpox monster crying out as it falls to an attack by people shooting arrows. Image courtesy of CDC/ Stafford Smith, Public Health Image Library (#2587).

leads, cultivating a climate in which people would sell their neighbor's health information for money.

Self-reporting and voluntary vaccination were encouraged in posters and pamphlets and even hand-scrawled messages painted on the backs of buses or banners slung over elephants. "Join the fight," encouraged one poster, as if recruiting troops for battle. It shows three Africans facing off with a smallpox monster. Marked with an identifying "SP" on its chest, the monster has distinct human features: arms, legs, fingers, toes, eyes, and hair. Its skin is black with white pockmarks, imitating smallpox lesions on dark human skin, and it holds a spiked club. It looms larger than the people pictured on the poster. The individuals facing the giant wield bows and arrows, one of which has been driven into the "heart" of the disease, squarely between the S and the P. The monster falls back, penetrated by the arrow, in an enactment of smallpox succumbing to the prick of a needle. The poster depicts a war against disease but personifies the virus enemy as a black monster. Infection demonizes humanity, positioning smallpox-infected bodies as the enemy of the healthy citizens who stand together and fight.

Other posters depict the vaccination act itself, a display of change enacted on a human body. In some posters, the injection gun is clearly displayed, poised to shoot a vaccine into exposed flesh of the victim's arm. Sometimes the person holding the gun is identified as a medical professional, wearing a stethoscope or a red cross; sometimes the gun is held by a uniformed officer, identified by a hat or badge. These posters illustrate the fundamental approach of the smallpox eradication program: one-on-one meetings between healthy people and government representatives bearing vaccine. The depictions establish a power structure in which the unvaccinated individual is under the control of the person administering the vaccine, underlining the vulnerable state of the unvaccinated body. Typically, the vaccine giver is a man, and the recipient is a woman, often a woman with a small child. The recipients often wear some form of "native" attire; the vaccinators wear uniforms. Though written as an invitation to be vaccinated, the poster affirms a larger system of social control enacted through the vaccination program.

Containment of smallpox required broad collective action. Quarantine has long been used to restrict the movement of disease, but it must often be enforced through authoritarian controls that impede powerful cultural rituals. During the WHO campaign, field officers reported traveling to the site of an outbreak to find scores of people traveling between towns to pay their respects to dead and dying relations, thereby carrying the virus back to their own homes.[39] In parts of India, the spring outbreak of smallpox was welcomed as the annual tribunal of the goddess Shitala Mata, by which she decided who was strong enough to live. Religious beliefs regarding animals, including the cow, created skepticism toward the vaccine, which was rumored to have bovine origins.[40]

Because ideas about disease integrate science, religion, and politics, they endow the medical subject with their own overdetermined values.[41] In reality, much of the smallpox campaign took place in communities with little exposure to modern medicine; the promise of the needle was

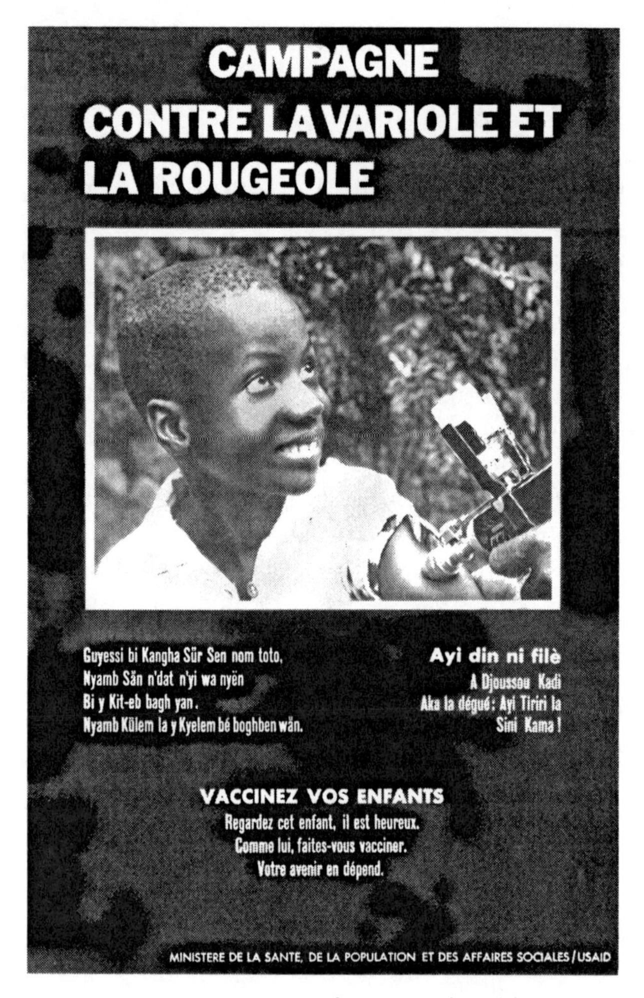

FIGURE 5 (a) Vaccination poster from West Africa delivering a message in three languages: "Look at this child, he is happy. Like him, get vaccinated. Your future depends on it." Image courtesy of CDC/Stafford Smith, Public Health Image Library (#2585).

hard to sell. Without the systems that had helped legitimate medicine in political decision making, vaccination had to be sold to communities by more direct strategies. "Often, before uneducated villagers would agree to submit, the vaccinators would have to jab themselves in front of the entire community, to prove that there was nothing harmful."[42] The propaganda promoting smallpox vaccination pulls on cultural values in order to push bodily acts. When the medical establishment fails to elicit

FIGURE 5*(continued)*. *(b)* Nigerian poster calling on citizens to fight smallpox. Image courtesy of CDC/Stafford Smith, Public Health Image Library (#2577).

desired behaviors from individuals, the broader political entity uses tools like mass media to elicit the social response that meets the necessities of governance. Thus medical practices integrate with other mechanisms of security by binding to broader communal values and community practice.[43]

The containment of smallpox also depended on identifying and distinguishing the immune population from potential carriers of disease. There are no invisible carriers of smallpox: people who are infected wear the symptoms on their bodies. People who are immune can prove their immunity by showing the scars on their arms from the vaccination or the scars on their bodies from the disease itself. WHO officials read these marks on human bodies to identify the "at-risk" population.[44] In later years of the campaign, officials traveled from house to house in search of the unvaccinated and infected.[45] On finding an active case of smallpox, they marked the house with a number, recorded the names of people living there, and then vaccinated everyone living within a quarter mile of an infected person, moving outward in concentric rings until they got ahead of the disease. The ring approach to vaccination shifted the campaign away from compulsory mass vaccination and used deviance as a basis for planning an intervention. By identifying people who were likely to become sick because of their proximity to an outbreak, the population could be understood in terms of probabilities rather than simply as "sick" or "not sick," and resources could be directed toward immunizing those most at risk. The ability to focus on the distribution of cases at a particular moment facilitated the "necessary and sufficient action of those who govern," a political response grounded in the assessment of risk and the basis of future bioterrorism interventions targeting vulnerable populations.[46]

Despite the radical idea that deliberate infection with a disease could create immunity, vaccination was commonly accepted because it was statistically successful. As the blight of smallpox began to fade, WHO officials scanned the planet for signs of any outbreak. Tentatively at first, they began to suggest the disease had been contained. In 1977, a man in Somalia became the last recorded person to catch smallpox from the body of another human being. His antibodies fought off the virus, and without another vulnerable body to infect, the disease succumbed. For two more years, a commission of scientists waited and watched, searching for signs of one more outbreak, before submitting a report stating that they found no mark of vital *Variola* on the world's population. The chain of contagion had been ruptured, and the ten-thousand-year-old virus no longer passed freely from host to host.

WHO president Abdul Rahman Al-Awadi pronounced smallpox dead on May 8, 1980, declaring "solemnly that the world and its peoples have won freedom from smallpox, which was the most devastating disease sweeping in epidemic form through many countries since earliest time."[47] The global economy, too, was freed from the toll of smallpox. WHO declared that the total cost of the campaign had been $112 million and predicted a worldwide savings of $1 billion annually through the eradication of smallpox.[48] The field workers were celebrated as the "heroes who conquered smallpox." At a celebratory parade in Sierra Leone, "the vaccination team members wore their field uniforms and displayed their jet injector guns for the public to see."[49] Such displays helped people to acknowledge a victory that was largely invisible to them, the elimination of a disease many had never known.

In the end, perhaps the most significant outcome of the WHO campaign was not the elimination of a disease but the worldwide expression of social control over a contagious disease. The metaphors of war and uprising, persisting to the very end in Al-Awadi's statement that people had "won freedom" from smallpox, established disease as an oppressive enemy that could and should be battled with all the weapons of modern medicine. The final victory shout of "Smallpox is dead" reaffirmed that the war had been against a living enemy and that success was to be marked by its obliteration from nature. Such approbation further ratified the practices of the campaign, justifying the body checks, quarantines, and martial law as displays of force necessary to protect the public's health.[50] Smallpox eradication was a celebration of science knowledge, the authority of developed nations over the third world, and the subjugation of individual bodies to achieve political and economic goals. It has been called a "symbol of human triumph over predatory nature" and a precedent for the eradication of other microbes.[51]

CULTURAL CONTAINMENT

Though the world was celebrating the "death" of smallpox, a number of *Variola major* organisms still existed in freezers in the United States and the Soviet Union. WHO's posteradication program called for all countries to send their laboratory stocks of smallpox virus to two repositories "to assure biosafety and security."[52] Scientists would have a ten-year window to study the virus through WHO-approved research projects, after which all remaining virus was to be destroyed. Beginning in 1993, a series of WHO resolutions delayed the destruction of the stockpiles on the grounds that advances in science and technology offered new mechanisms for studying the virus and gaining new disease knowledge.[53] WHO continues to deliberate the risk and benefit associated with the smallpox stockpiles.[54] The reluctance to destroy a germ—even the most deadly germ in human history—reflects the belief in the power of scientific study to produce benefits that outweigh even the most substantial risks.

The eradication of naturally occurring smallpox heightened its interest as a bioterror weapon. The possibility that it lives on, in places unknown to WHO officials, also shapes the contemporary assessment of the bioterror threat. WHO acknowledged this possibility in a 2007 resolution "recognizing that unknown stocks of live variola virus might exist, and that the deliberate or accidental release of any smallpox viruses would be a catastrophic event for the global community."[55] Some researchers estimate that up to seventeen countries are holding live virus for offensive or defensive research purposes.[56] Policy makers are no longer willing to base stockpile-management decisions on the assumption that smallpox is securely contained in two freezers.

Moreover, advances in genomic biology raise the possibility that smallpox will live forever. In 2002, scientists at a New York university created a polio virus from small fragments of its DNA sequence. Although the smallpox virus is much more complex (185,000 base pairs to polio's 7,000), that project shows the potential for synthesizing viruses.[57] Much of the smallpox DNA sequence has become publicly available, as research institutions have worked with the virus to develop vaccines and antidotes. In 2006, a *Guardian* reporter ordered a smallpox sequence over the Internet, using a fake company name, a cell phone, and a residential address.[58]

Others worry that the virus might be lying dormant in cemeteries or glaciers, set to reemerge in nature as global temperatures climb. In 2003, a librarian in New Mexico discovered smallpox scabs in an envelope inside a library book. Though they contained no living virus, the genetic technologies of the twenty-first century might allow the creation of live virus from such material. To public knowledge, smallpox does not exist in nature, but the disease continues to wield political force through its imagined potential to emerge and infect a newly vulnerable population. The threat of disease can now be understood both as an unpredictable and uncontainable form of nature and also through the fear that humans are accumulating greater ability to harness microbial nature and use it to odious effects.

The work in the 1960s and 1970s to eliminate *Variola* seems contrary to parallel political movements in the United States to save endangered species. Some arguments against destroying the smallpox stocks center on environmental law, biodiversity, and the rights of nature. Species deemed worthless at one point in history may be assigned different cultural values at a different time: wolves, for example, were extirpated from the U.S. West and later restored—at some expense—for ecological benefit. The Convention on Biodiversity mandates the preservation of genetic material, and smallpox is a very old, well-adapted organism whose genetics may be revealing. Some evoke the precautionary principle, so often applied to environmental actions, to require those who would irrevocably destroy the virus to demonstrate that no unintended negative consequences will result from the annihilation of the organism. Yet the environmental standards by which we assign cultural value to nature fall short of accounting for a species that has caused so much human harm. Some who wish to destroy the stockpiles point to the horrible social consequences of the disease or imagine its use in a bioterror attack on a newly vulnerable population.

Because smallpox still exists, albeit under rigorously controlled circumstances, the microbe continues to set the terms of human vulnerability in the modern age. After the Smallpox Eradication Programme, medical professionals were discouraged (and at times prohibited) from vaccinating individuals against smallpox on the grounds that the risk of illness from the vaccine posed a greater threat than the disease. Thus an entire generation—in some countries two or three generations—has not been vaccinated against smallpox, and even those who bear the scar of the vaccine have lowered immunity today. The power of smallpox to shape current biosecurity practices draws on the cultural understanding of smallpox as a brutal, deadly disease as well as on the political production of a population made vulnerable, ironically, through the elimination of the threat.

AN ARMY TO FIGHT A KILLER VIRUS

"A virus that kills every one of its victims," opened an article in *New Scientist* in 2001. Under the headline "Killer Virus," the reporter Rachel Nowak disclosed how two Australian scientists had "accidentally" created a lethal mouse pox virus and intended to publish their findings in an academic journal. "How do you stop terrorists taking legitimate research and adapting it for their own nefarious purposes?" she asked. In response, the article quoted D. A. Henderson, then with the Johns Hopkins Center for Civilian Biodefense Studies: "I can't for the life of me figure out how we are going to deal with this."[59] Media outlets developed dystopian fantasies of scientists concocting deadly superbugs through genetic engineering.[60] Months before the anthrax attacks in the United States, Nowak's report stimulated the cultural debate over how genetic modification might change the landscape of disease. If scientists could increase the virulence of a pathogen, disease could no longer be understood as bounded by the laws of nature. The very terms of vulnerability had to be recalculated in light of this new biological threat. The events that followed show the particular role of the science industry in defining risk and the controls placed on the flows of information in the name of national security.

The day after Nowak published her exposé, and a month before the publication of his findings in an academic journal, the Australian scientist Ron Jackson heard a television news report that he had "re-created smallpox" and built the "ultimate weapon" in his laboratory. Though he had noted the implications of his research early on, Jackson marveled at the ominous twist the media put on it.[61] Essentially a well-equipped exterminator at the Pest Animal Control Cooperative Research Centre in Canberra, Australia, Jackson was exploring a theory that he could create "contagious sterility" by fusing a virus with proteins from a female mouse's unfertilized eggs, causing the mouse's immune system to kill her own eggs.[62] Jackson and his partner from the Australian National University, Ian Ramshaw, spliced a mouse pox virus with an interleukin-4 (IL-4) gene to enable it to immobilize the healthy immune system of lab mice. They injected the test virus into ten mice, and after six days, one of them was not only sterile but dead; two days later, three more died. Eventually, all ten mice died from the injection, offering evidence that the process dramatically increased the virulence of the virus.

Jackson and Ramshaw had been working with the poxvirus strain that causes mouse pox. Like smallpox, this virus is highly contagious, but its spread can be limited by a vaccine.[63] In the next research phase, they first immunized the mice against the mouse pox virus and then injected them with the same combination of virus and IL-4 protein. Despite immunization, six of the ten mice still died from the injection. This was the result that, according to one reporter, "fundamentally altered the world's terror equation."[64] Jackson and Ramshaw had found a way to defeat the power of vaccination. The experiment potentially undermined a political system that relied on cheap and highly effective vaccines to secure bodies from contagious disease. It also created a new reality in which humans can engineer viruses that can defeat the vaccines tailored to resist infection. Back in that Australian laboratory, though, the scientists had dead mice and a dilemma: should they publish their results, thus publicly disseminating instructions for creating such hypervirulence?

The handling of these findings by the scientists and their institution exemplifies the ethical debate in the science community over the dual-use potential of new biotechnologies and the publication of results that might provide a blueprint for building bioweapons. Jackson and Ramshaw consulted their peers and instigated an inquiry that eventually reached the Australian Department of Defence.[65] All the organizations said they should publish. Believing that open discussion of the topic would best serve the public, the scientists submitted an article to the *Journal of Virology*. They made no reference to the weapons potential of their findings, and none of the scientists who reviewed the article commented on it. A month before the journal article was published, however, Nowak's article set new terms for the debate.[66]

The Australian mouse pox experiments caught the media's fancy for a number of reasons, including a waxing public interest in bioterrorism. The media told a story of bumbling scientists who slipped up in the lab, produced a deadly disease, and then fearfully withheld their data because of its implications. However, transgenic microbe research had been carried out for nearly three decades, and the suggestion that IL-4 could be used to make a virus stronger had been circulating in science publications for about five years. What distinguished Jackson and Ramshaw's research from all these former experiments was its focus on mouse pox.[67] While mouse pox is harmless to humans, the related *Variola* virus coevolved with humans to become the most deadly virus in human history. Jackson and Ramshaw's experiment raised alarm not because it killed mice so effectively but because the virus with which they worked was related to a disease associated with death and horror in human populations. The cultural, economic, and political experiences that have for centuries given meaning to disease continue to shape people's understanding of the threat of bioterrorism.[68] Imagining germs as instruments of war and terror will continue to circulate and add to these meanings.

The threat of disease as an offensive weapon leads to interventions in the bodies of soldiers. Noting the potential for smallpox to be used as a weapon, General George Washington ordered that his troops take measures to protect themselves against the disease.[69] Fifteen months after the September 11 terrorist attacks on the United States, President George W. Bush announced a plan to vaccinate one million military and public health personnel against smallpox. Hours before vaccinations began at Walter Reed Army Hospital, the president delivered a speech describing a world where smallpox officially did not exist but paradoxically still posed enough of a threat to justify taking action on the bodies of soldiers:

> In 1980, the World Health Organization declared that smallpox had been completely eradicated, and since then, there has not been a single natural case of the disease anywhere in the world. We know, however, that the smallpox virus still exists in laboratories, and we believe that regimes hostile to the United States may possess this dangerous virus. To protect our citizens in the aftermath of September the 11th, we are evaluating old threats in a new light. Our Government has no information that a smallpox attack is imminent. Yet it is prudent to prepare for the possibility that terrorists [who] kill indiscriminately would use diseases as a weapon.[70]

In this speech, the known presence of smallpox in two laboratories (one of which is located in the United States) and a belief that it may exist elsewhere are invoked as justification for reviving discussion of an "old threat." The memory of a disease with no cure, which kills 30 percent of its victims, still produces enough fear to rationalize government intervention.

Bush's vaccination plan reproduced forms of social control used in the global smallpox eradication campaign, forming a ring of immunity by targeting first the bodies of military and public health workers and then emergency responders. Although mass vaccination was declared unnecessary, the vaccination of military and emergency personnel was justified "for the greater public good" and because "military missions must go on even if a smallpox outbreak occurs[, and] vaccination is a wise course for preparedness and may serve as a deterrent."[71] By vaccinating soldiers, the government was not only prioritizing the vitality of the military but also producing immunized soldiers to deter enemies' use of smallpox as a weapon. Soldiers impervious to disease were an asset of war.[72]

Vaccination, however, carries a risk. In the face of an outbreak of smallpox with a 30 percent mortality rate, this risk is relatively small, but when the chances of contracting the disease are close to none, two deaths per one million vaccinations is a considerable risk. President Bush declared that he would be vaccinated: "As Commander in Chief, I do not believe I can ask others to accept this risk unless I am willing to do the same. Therefore I will receive the vaccine along with our military."[73] He added that his family and staff would not be vaccinated, exemplifying the appropriate response for ordinary citizens.[74] Tommy Thompson, secretary of health and human services, explained that the government would control distribution of the vaccine: "It will not be in your doctor's office. We will not give it up out of our custody. It will not be willy-nilly handed out to doctors across America. We will retain custody of the vaccine."[75] Control of the remedy was critical to retaining the state's authority over disease.

In the end, not even phase 1 of the vaccination plan, the inoculation of one million bodies, was completed. Health care workers were unwilling to accept the risk, and only thirty-eight thousand volunteered for vaccination. Two individuals died of heart failure after receiving the vaccine. As public concern about smallpox waned, some argued that the directive had already done its work, drumming up support for the war in Iraq.[76] Attention shifted from immediate mass vaccination to stockpiling vaccines and planning for their distribution during a state of emergency. Within a year of the terrorist attacks in 2001, the United States had purchased enough smallpox vaccine to inoculate every American citizen. The timely distribution of the vaccine following an outbreak would be as effective as and less risky than mass vaccination.

Modern bioterrorism preparedness plans have drawn explicitly from the preemptive preparedness strategies of the Cold War, which created subjects for the national security state by producing vulnerable bodies. Today, the government stockpiles vaccines alongside its warheads, ready to defend the homeland. Vaccination remains the nation's first defense against smallpox, even as scientists present growing evidence that vaccination can be overcome by modern microbes. Techno-scientific fears once reserved for nuclear bombs now extend to mutant ecologies.[77] If the human body is the terrain of modern warfare, its state of immunity must be disciplined to eliminate fear and vulnerability.[78]

A COLD WAR ON TERROR

"'Your body is under constant attack!' Give your immune system the support it needs. Order Now!" Dr. Ken Alibek's Immune System Support Formula offers an "Advanced Natural Health Formula" in capsule form, promising to infuse the body with a tide of vitamins, minerals, antioxidants, and probiotics designed to boost immunity. Alibek's website declares that "$1.34 a day is a small price to pay to support your immune system," assigning a monetary value to health. Borrowing language from the security industry, the site describes the immune system as "your body's personal surveillance and security system[, which] recognizes foreign substances that are constantly attacking your body and it helps defend against them."[79] Guardians of personal health, it is assumed, will respond by readying the body to fight biological threats.[80] Indeed, the conception of a body under attack from outside and within characterizes the biological subject of the national security state.[81] The duty of the biological subject is to attend to the vulnerabilities within the body and in its physical surroundings and to mitigate risk. In the emerging culture of biological insecurity, a body constantly under threat is as a microcosm of the nation-state.

Dr. Alibek, the purveyor of healthy immune systems, is a self-described "biological and medical expert . . . internationally recognized for his groundbreaking research on the human immune system, . . . consultant to the U.S. government," a contributor "to world peace" and "a U.S. citizen."[82] While Alibek's credentials as a natural healer may be questionable, this former head of the Soviet Union's Biopreparat program gained fame by exposing the Soviets' secret bioweapons development, which continued after international treaties prohibiting biological warfare were signed in the 1970s. Between 1998 and 2005, Alibek testified before Congress at least ten times, presenting stirring, firsthand accounts of the Soviet Union's plans to genetically engineer pathogens under projects code-named Bonfire and Podleshik, painting a vivid depiction of a foreign threat.[83]

Alibek's willingness to talk about Soviet weapons development facilitated the United States government's effort to establish bioterrorism as a credible threat. His testimony enabled a broad field of nations and groups to be identified as potential threats to the United States, and his Russian-accented words brought the memory of the Cold War to bear on the contemporary bioterror crisis. Both Alibek's magic immune pills and his revelations of Soviet secrets reconfigured the fears of the previous century. To understand how these ideologies coalesce with cultural histories of disease, it is useful to investigate conceptions of life and state in the politics of the mid-twentieth century.

The Cold War was about the defense of the nation-state, but the production of the state is simultaneously and insistently the re-creation of its citizens. As Judith Butler suggests, discord between the state and the individual is an inevitable outcome of their relationship: "The state then makes us out of sorts, to be sure, if not destitute and enraged. Which is why it makes sense to see that at the core of this 'state' . . . [are] juridical and military complexes that govern how and where we may move, associate, work, and speak. If the state is what 'binds,' it is also clearly what can and does unbind."[84] The Cold War might be understood through the ways people negotiated tensions between the nation-state and individual states of being in the presence of nuclear weapons; it was about how people conceptualize mass death at home and abroad.[85]

As nuclear weapons remade the world around the possibility of instantaneous mass death, they changed the nature of life itself. Joseph Masco argues that the nuclear project of the Cold War brought the possibility of sudden death into formations of healthy citizenship. Security and health became "contradictory ideas" during the nuclear era, and this development "both underscored the reality of radical technological change and invalidated the state's ability to regulate society at the level of health and happiness."[86] Unable to build the nuclear industry while caring for its citizens' health, the government instead worked to naturalize the atomic threat as one for which citizens themselves must prepare. Relieved of the burden of protecting individual health, the state could focus on survival as the preeminent goal of citizenship.[87]

Although the nuclear threat has not disappeared in the new millennium, the remaking and potential weaponization of pathogens through genomic science has produced new natures that are seen as vital security risks. The case of smallpox shows how hybrid nature-cultures may create unending, unspecified risks. Situating the production of biological risk within broader security narratives of the Cold War lays a foundation for understanding the emergence of biosecurity as a means of governance in the post-9/11 world. Practices developed to counter nuclear threats during the Cold War have been remade by an all-hazards preparedness strategy sustained in part by the production of living, changing, mutating, and growing biological threats. The creation of the Department of Homeland Security appropriates the consolidation of executive power, the militarization of civilian life, and the expansion of the science industry that began during the Cold War to manage today's threats.

Consolidation of Executive Power

During World War II the United States greatly expanded its systems of governance in order to support U.S. interests far beyond its borders.[88] In the following years, without a war to sustain the power of the central-state, the country seemed ready to revert to the decentralized policies that invigorated American capitalism. To avert this outcome, the consolidated government powers produced an "imaginary war."[89] The preemptive war was fought daily through practices that promised to eliminate vulnerability but reached deep into the public and private lives of citizens. To wage the Cold War, government had to strengthen the power of the executive while still upholding the democratic social order, convincing citizens to accept the expansion of government as an innocuous consequence of work for the public good.[90]

In the wake of the bombings of Hiroshima and Nagasaki, worldwide nuclear war seemed possible, even probable. Nuclear war was a risk not only to bodies and life but to the balance of power that preserved life. Ulrich Beck argues that principles of risk work with material objects, like a bomb, to order the world and render it governable. His thesis rests on the idea that risk exists beyond society's ability to calculate it, for the ultimate products of modernization are catastrophic events whose effects cannot be fully measured, such as the nuclear meltdown at Chernobyl. Similarly, pathogens hold the possibility for contagion, mutation, and rapid adaptation, with potential effects that are incalculable.[91] The ordering of events in the cultural narrative of smallpox exemplifies how vaccination worked to make disease governable: it transformed likely death into a calculus of risk and then opened the possibility of partnering science with politics to mitigate risk. Thus the disease risk is technologized by practices in the present that work to mitigate future loss.[92]

The national security state operates under the assumption that, given the necessary authority, resources, and access, risk can be controlled, thereby validating the expansion of governance to acquire the means to mitigate risk. The "red button" of the nuclear era imagined an executive who could decide the fate of millions with the touch of a finger. Living in a state where one person held such power demanded that citizens trust their political leaders. Throughout the Cold War, the executive branch of government retained extensive authority over risk control. In a nation built on the principle of distributing executive powers to individual states, federal agencies claimed startling access to the lives of citizens in the name of civil defense. Many events of the contemporary era, from the Patriot Act to the establishment of the Department of Homeland Security, mirror the Cold War bestowal of executive power in the national security state.

The Militarization of Civilian Life

Even in its infancy, national civil defense policy assumed citizen participation and citizen responsibility. Though states accumulate power, they do not control all power. Even during a crisis, citizens retain the power to act or not.[93] Because civil defense grew to be increasingly expensive, unprincipled, and dangerous, the public had to be morally persuaded to pay any cost for freedom. Accepting that level of sacrifice required a conviction that life as they knew it would continue after an attack.[94] The goal of civil defense was to convince the public that they could survive a nuclear attack, rebuild, and return to their everyday lives. Under direction from the Commander in Chief, people dug bomb shelters, stockpiled food, and ran drills at home and at work. As the country slipped into a state of war, the civic accepted the military rituals and practices that pervaded their daily lives. From the perspective of the state, nuclear bombs were not the enemy: panic was the enemy. The national security campaign encouraged the rehearsal of nuclear fear to stave off mass panic while keeping the threat ever-present in the public mind.

Part of the Cold War project involved producing the future as if it had already happened.[95] It had to be a place where people still lived familiar lives. Unable to change the future, therefore, people had to rearrange their present lives to conform with imagined postnuclear life. Preparedness thus brought future peril into a space where it could be acted on in the present.[96] When the practices of crisis become the practices of everyday life, the "distinction between crisis and normality" collapses.[97] The Cold War might therefore be understood as a moment when the public accepted living with crisis as the normal way of being. The preparedness rituals that reset the parameters of life during the perpetual state of emergency reemerged with the war on terror.

The Expansion of the Science Industry

Notably, civil defense could be practiced at home by any citizen (though, if one takes the characters in defense films, advertisements, and pamphlets as examples, those who did so were likely white and middle class), but the nuclear problem was a technological problem and therefore one in need of a scientific fix. Colin Gray suggests that the appeal to science resonated with the "American engineering spirit."[98] Scientific modeling informed the public about the range, impact, and duration of a nuclear blast. The postwar era also saw the expansion of an offensive bioweapons program, an endeavor that imagined the possibilities of transforming biological agents into as-yet-unidentified superweapons. As the science industry grew, it became yet another expansion of state power. The increased role of science in risk management instigated broader discussions of the power of science over human lives.[99]

François Ewald contends that the Cold War love affair with science cannot endure the new paradigm of security. The rapidly evolving science of today, constantly swept on to the next question or catastrophe, is losing its ability to mitigate cultural fears and insecurities. "While the language of risk, against a backdrop of scientific expertise, used to be sufficient to describe all types of insecurity, the new paradigm sees uncertainty reappear in the light of even newer science. It bears witness to a deeply disturbed relationship with a science that is consulted less for the knowledge it offers than for the doubt it insinuates."[100] When scientists began looking at microbes through simple microscopes, they constituted disease as an entity external to the human body and therefore a risk that could be managed through technology and social practice. In the modern era, when scientists can change the materiality of microbes in a way that rattles our assumptions about the nature of life itself, as Jackson and Ramshaw did in their laboratory, uncertainty grows. In that moment, political actors like President Bush turn to science in search of doubt, for uncertainty incites scientific action and rationalizes the expansion of governance.

BIOSECURITY AT HOME

The Homeland Security Act of 2002 revived Cold War concepts of civil defense and established a new government agency charged with preventing and responding to terrorist acts. The Department of Homeland Security (DHS) models elements of the Cold War state by consolidating authority, strengthening the power of the executive branch to respond to the threat of terrorism, and enlisting citizen participation in securing everyday life.[101] The mission of the new department was legislated as follows:

> (1) IN GENERAL.—The primary mission of the Department is to—
> (A) prevent terrorist attacks within the United States;
> (B) reduce the vulnerability of the United States to terrorism;
> (C) minimize the damage, and assist in the recovery, from terrorist attacks that do occur within the United States; . . .
> (F) ensure that the overall economic security of the United States is not diminished by efforts, activities, and programs aimed at securing the homeland.[102]

References to "prevention," "vulnerability," and "securing the homeland" in these lines echo the Cold War ideology of deterrence and survival. The mandate to "assist in recovery" assumes survival and the continuation of life, while ensuring "economic security" assures that these executive efforts will not interfere with American capitalism. Notably, the DHS is an agency that operates because of terrorism threats, but not exclusively during terrorist events, for the executive is continually working to reduce vulnerability and rehearse the terrorism response.

The DHS assembled twenty-two government agencies in a unified mission to achieve domestic preparedness. This reorganization, the largest restructuring of U.S. government since President Harry Truman combined the war and navy departments into the Department of Defense in 1947, brought agencies ranging from the Secret Service to the Coast Guard to the Immigration and Naturalization Service into a new cabinet-level department.[103] Under U.S. law, departments like DHS are both "an executive agency and a military department," a classification specifically referenced in the Homeland Security Act (USC 101, Intro, Sec 2.7). Through this reorganization, agencies concerned with citizenship, economy, and security were intertwined and given responsibilities to protect the homeland.[104]

The first directive of the new DHS was "to carry out comprehensive assessments of the vulnerabilities of the key resources and critical infrastructure of the United States, including the performance of risk assessments to determine the risks posed by particular types of terrorist attacks within the United States," thus beginning a formal calculation of risk and assessment of vulnerability by the new security state.[105] The DHS definition of critical infrastructure comes from the Patriot Act (section 1016(e) of Public Law 107–56), including "systems and assets, whether physical or virtual."[106] Vitality is measured not only in terms of survival and public health but in the continuance of the national economy: "The term 'key resources' means publicly or privately controlled resources essential to the minimal operations of the economy and government."[107] This language enables an expansive response to terrorism preparedness, in which federal authority manages both the health and the wealth of the nation in an ongoing state of emergency.

The establishment of an ever-present "war on terror" creates a climate that renders citizens more willing to surrender civil liberties to the state. As the DHS strategic plan explains, "This exceedingly complex mission requires a focused effort from our entire society if we are to be successful."[108] In the introduction to its Ready.gov program, DHS issued the following statement, now cited word for word on more than one hundred local-authority preparedness websites:

> All Americans should begin a process of learning about potential threats so we are better prepared to react during an attack. While there is no way to

> predict what will happen, or what your personal circumstances will be, there are simple things you can do now to prepare yourself and your loved ones.[109]

Citizens are then invited to take specific actions, like assembling an emergency supply kit and developing a family emergency plan, because "preparing makes sense." Again, preparedness is situated as a personal responsibility overseen by the state.

Though the narrative of national security produced since the September 11 attacks mirrors Cold War civil defense strategies in several ways, the current situation poses two different challenges. First, the Cold War had an identified enemy in the Soviet Union, while the war on terror involves unknown enemies, who may be individuals or well-organized groups. To produce the requisite levels of fear to instigate social mobilization, the state must convince the public that terrorists pose a significant threat. Yet if the enemy continues to be unidentifiable, the risk may, as Beck suggests, slide into the realm of the incalculable, rendering the state's deterrence programs inefficient and ineffective. As Priscilla Wald observes, the transmission of disease, which defies boundaries, troubles the effort to identify a particular nation as the monolithic enemy of the war on terror but may prove effective in reifying national borders through disease deterrence programs.[110]

A second problem for the contemporary security state is the absence of a single, identifiable weapon. During the Cold War, the threat of the nuclear bomb gave the state a way to access citizens' fears and control their responses.[111] The bomb was deadly enough to pose a far-reaching threat to human life but simultaneously appeared controllable: it was contained in a silo, controlled by that red button; it would detonate in a single blast; and according to the science of the time, it would have limited long-term effects. The modern U.S. security state grouped biological threats with chemical and nuclear bombs under the label of weapons of mass destruction (WMD) because it needed to elucidate a singular threat analogous to the nuclear bomb. Biological weapons can be presented as a threat that can be made by one person on a small scale but has vast potential to spread, reproduce, and mutate. Furthermore, just as the threat of the nuclear bomb corresponded to the resources and political power of the Soviet Union, biological weapons match the resources of terrorists, for they emerge from nature itself, a resource available even to underfunded terrorists. As one vaccinologist noted, "As the might of the U.S. increases and the poverty of other nations increases . . . what weapon do they have to strike back with? The only one they can afford and the only one we might not be protected against are biological weapons."[112]

Not only are bioweapons broadly perceived to be cheap, available, and easy to deploy, but they perpetuate individual vulnerability. The testimony of Roger Brent, director of the Molecular Sciences Institute, before the House Subcommittee on Prevention of Nuclear and Biological Attack exemplifies the expansive response seen as necessary to prepare for biological events: "Because this threat has changed from the days of the Cold War germ war program, our defense posture needs to change. Although it is a good thing we now have enough smallpox vaccine . . . it is important to remember that stockpiles of vaccines and drugs are fixed defenses against known threats."[113] Preparing for the unknown demands a continual anticipation of new threats. Brent argued for uncontested support of this national science project: "Building a defense is a problem of real gravity and complexity; it will require R&D and policy efforts sustained over decades, which will mean that it will need to enjoy sustained consensus bipartisan support, as was true for Government support for science and technology during the Cold War."[114] This direct reference to the nuclear era in support of expansive science situates government and science as cocreators of national security, a role formalized in the Science and Technology Directorate of DHS, whose mission statement brandishes science and technology as tools to "enhance security and increase efficiency."[115]

The direct links between science and civil defense motivate technological fixes for security threats and drive the development of technologies that may not be scrutinized for their cultural value or their implications for civil liberties. In the words of President Bush, "We refuse to remain idle while modern technology might be turned against us; we will rally the great promise of American science and innovation to confront the greatest danger of our time."[116] When this danger includes incalculable biological threats, it is a patriotic necessity to produce equal or greater innovation in the homeland.

In its concern with protecting critical infrastructure, the contemporary national security state must now deal with individual bodies as never before. A bioattack will not destroy buildings, but it may still dismantle social systems. The Cold War civil defense narrative imagined a state of survival in which cities were destroyed, but people who were prepared for attack survived to rebuild; bioterrorism operates within and between bodies, spreading silently within an intact infrastructure to destroy life. In his testimony, Brent suggested that mechanical weapons could be removed from the scenario completely, for terrorists could infect themselves with disease in order to infect other people. The committee chair, Christopher Cox, explicitly alluded to the limitations of traditional security practices in his response to Brent's suggestion of "suicide coughers":

> *Mr. Cox:* And what you suggest, therefore, is that the Cold War model, or really the model of all prior history in warfare is out the window; we shouldn't be looking necessarily for weaponization, the terrorists themselves become the weapons. Is that what you are suggesting?
>
> *Dr. Brent:* That is correct, sir.[117]

In this world of biological risk, the national security program treats bodies as both casualties and weapons of mass destruction, and as both victims and vectors of disease. A biological attack can continue as long as vulnerable bodies exist. Biological weapons demand new knowledge of disease and rationalize the expansion of science research.

SCARS

Variola major had done powerful cultural work by the time the virus was seen through a microscope around the beginning of the twentieth century. Those first glimpses were of a form commonly described as "brick-shaped," with its DNA contained in a dumbbell-shaped membrane at its core. The material shape of the virus gave insight into how it works, latching onto a host cell within the human body and then releasing its core protein into the cell to replicate and create new virions. This process can be understood as natural, the work of the virus to reproduce and ensure the survival of its DNA, even as the manifestations of those unseen events on the human body bring rapid cultural effect. The work of the virus within the body appears determined, but even the microscopic world contains uncertainties, such as the effects of antibodies that impede the virus's reproductive work. Thus, smallpox exemplifies the eventful nature of "things," for the virus is material and meaningful, but it also carries an uncertain future, "a margin of indeterminancy."[118] For millennia smallpox viruses moved through the world, interacting with bodies, communities, governments and institutions. Today they still act as a creative force despite—or because of?—their deeply frozen state. The work of smallpox extends beyond its material reproduction; it is part of a political ecology of things, working to shape the institutions of the modern world. Smallpox assembles our social future even as it subjects us to a past we did not create.[119]

As such, the virus is an actant, in the Latourian sense, or a "small agency," following Charles Darwin. The work of viruses has shaped human history as an unplanned effect of their own exertions to survive. As Jane Bennett explains, these small agencies "participate in heterogeneous assemblages in which agency has no single locus, no mastermind, but is distributed across a swarm of various and variegated vibrant materialities."[120] The virus is part of a system that is both political and ecological. Following John Dewey's theorization of the public as bodies brought together around a shared problem, Bennett imagines a political ecology in which nonhuman and human collectives brought into existence through the shared experience of harm create democratic future paths through the acknowledgment of vital nonhuman agents that cocreate the world. The pathogenic, nanoscale actants affect humans and other organisms through disease, but their work is also a set of events whose outcomes are not determined and which open the possibility of many futures. Embracing smallpox and other germs as vibrant materialities breaks apart the structures of power and politics that have defined the bioterrorism crisis. The publics created around the shared experience of disease must be responsive to vibrant nonhuman agencies that cocreate the world, bringing them along in the work to build more-just futures.

The materiality of the microbe cannot be separated from the physical manifestation of infection in the human body. Disease symptoms express the presence of the nonhuman, creating a human-pathogen hybrid that has its own materiality and politics. The materiality of *Variola major* is expressed in pustules, scabs, and scars. In the political system, the occurrence of smallpox can be read on human bodies, bodies that are also simultaneously read for race, gender, age, and geography. Disease and health become categories of difference. These bodily discourses include a range of hybrid states and exist on many scales, from those that can only be seen through science practice, such as a laboratory test, to those that can be observed walking down the street, such as smallpox scars. Immunity and health can be read on bodies and calculated in the assemblage of the healthy body. Take, for example, a case in which the manifestation of acts to create immunity within human subjects enters into political discourses of race, immigration, and national security.

In the last moments of a 2005 Congressional hearing on engineering bioterror agents, Representative John Linder asked a panel of bioterror experts: "What would you say if I told you a scientist from Sweden said that Iranian children emigrating with their parents from Iran to Sweden have all been vaccinated for smallpox; what would that mean to you?"[121] A topic so ripe with hearsay and speculation was troubling to scientists, and the hint at secret government knowledge was enough to create more than a murmur in online communities.[122] In itself, however, the question exemplifies the ongoing effort to make bioterrorism visible in the contemporary moment, in this case by reading human bodies. The cultural history of smallpox that shapes knowledge of modern bioterrorism gives a scar political meaning. In Linder's scenario, the scarred bodies emerge from a cultural geography that labels them as Middle Eastern, Iranian, emigrant, and foreign, and they are read, first by a European scientist and then an American politician. The scar does not render these bodies diseased but rather the opposite. The speaker is asking whether a marker of smallpox immunity manifest on subjects' bodies can be read as evidence of state terrorism. Cultural fears of Iranian terrorists and smallpox converge on the body of a child with a scar.

Sitting on the panel of expert witnesses, Ken Alibek suggested that people in Iran simply are not convinced that smallpox has been eradicated and so continue the vaccination practices of previous generations. He argued that the scars express a social fear of disease, evidence of the lingering effects of the world's deadliest disease in a developing nation. Linder's question evoked multiple narratives that might make sense of these scarred bodies, while strongly hinting at the answer that best served national security objectives. In asking whether the fears of Iranian emigrants should be shared by Americans, the question scrutinizes migrant bodies in order to understand risk. Racialized notions of risk attach to national security through these bodies. They evoke narratives that have reverberated throughout the history of smallpox and are now echoed in narratives of terror. The scars, imagined or not, keep smallpox bound to the present moment, producing a narrative of risk ridden with secrecy, foreign bodies, migration, health, and vulnerability.

The threat of bioterrorism demands the reconfiguration of national security practices based on fear and knowledge of disease. It should not be assumed that "new" knowledge of disease will eliminate fear, for, as with smallpox, even the eradication of disease does not eliminate culturally constructed vulnerability. When knowledge itself is rendered suspect, Masco argues, "only ideology and desire remain as the basis for action."[123] As citizens project their memory of smallpox onto the blank screen of the future, they imagine risk according to deeply personal conceptions of life and fear as well as the collective experience of disease brought into public discourse.[124] At the center of this paradigm lies the objective of survival, the protection of life above all other social institutions and civil liberties.

The forms of governance that attend to everyday life enter vigorously into twenty-first-century society, responding to the variety of threats produced by the conflated histories of disease and national security. Because bioterrorism aligns closely with the history of infectious disease, the politics of smallpox illuminates the origins of future bioterror risk in the experiences of the deadly diseases of the past. The smallpox eradication campaign exemplifies how disease control practices harness cultural institutions to achieve state objectives during a war on germs. Jackson and Ramshaw's publishing dilemma shows how the fear of disease and bioterrorism disrupts the scientific production of knowledge, evoking the culture of secrecy that shaped Cold War society. The DHS has been built around similar cultural fears that consolidate the power and authority of government and the science industry over people's lives. Even a cheap vitamin pill, guaranteed by an internationally renowned bioterrorism expert to protect one's immune system, shows how individual pursuit of security against a natural world bursting with pathogens enters the market, aligning personal health behaviors with the concerns of the security state. Cultural fears are complex, and fears of bioterrorism are no exception. Addressing these fears will demand broad reconsideration of how people understand their bodies and how living entities, large and small, operate within the security state.

2

Microbes for War and Peace

On the Military Origins of Containment

Disease and war have a long alliance. In the United States, the modern bioterror lab came into being when the army transformed a decommissioned airstrip in rural Maryland into a top-secret facility to study the potential for building biological weapons. The United States was late to join the field of biological weapons research: Japan, Germany, Great Britain, and Canada had been experimenting with biological agents in the interwar period and throughout the Second World War.[1] The place of germs in modern warfare was highly debatable, though the Geneva Protocol in 1925 included "bacteriological methods of warfare" along with poisonous gases as forms of warfare that had been "justly condemned by the general opinion of the civilized world."[2] The protocol hints at some international consensus on the morality of biowarfare, but there was little understanding of what a technological bioweapon would look like. In 1941, the U.S. secretary of war, Henry L. Stimson, asked the National Academy of Sciences to investigate the potential of biological warfare. Stimson chartered the War Bureau of Consultants, a committee of scientists who began a systematic study to determine whether the potential for biological weapons conjectured by microbiologists was indeed feasible:

> The value of biological warfare will be a debatable question until it has been clearly proved or disproved by experience. The wide assumption is that any method that appears to offer advantages to a nation at war will be vigorously employed by that nation. There is but one logical course to pursue,

FIGURE 6. In a mid-1950s field trial at Porton Down, UK, participants wore masks to collect the aerosolized substances dispersed by airplanes. Early bioweapons research mapped new geographies for microbes moving through air, which also influenced containment practices. Image courtesy of Imperial War Museum, London.

> namely, to study the possibilities of such warfare from every angle, make every preparation for reducing its effectiveness, and thereby reduce the likelihood of its use.[3]

Scientific study promised to both define and eliminate the potential for biological warfare. Public interest and government resources brought in through military funding streams would jumpstart the emerging disciplines of bacteriology and microbiology. That countries like Great Britain already had bioweapons research programs further fueled interest on the part of the United States.

On the recommendation of the National Academy of Sciences, President Franklin D. Roosevelt established the War Research Service, a civilian agency to coordinate research in government institutions and the private sector. Under the direction of the pharmaceutical mogul George W. Merck, the War Research Service sponsored studies at dozens of universities and private industries, particularly systematic surveys by experts of microorganisms deemed to have weapons potential. Within a year, however, the agency recommended that a facility be created to scale up the research program, and work began to transform the army's Camp Detrick in Frederick, Maryland, into the hub of the U.S. bioweapons program. In the climate of war and amid rumors that Japan and Germany were planning to deploy biological weapons, funding for the War Research Service doubled in 1944, to $400,000. In addition to the Maryland research facility, planning began for a biological weapons production facility in Indiana and testing grounds in Mississippi and Utah. In June 1944, Roosevelt dissolved the civilian War Research Service and put the biological weapons program under the authority of the War Department. Merck, now a special consultant on biological warfare to the secretary of war and chair of the United States Biological Warfare Committee, reported in 1945 on the work done during the war, emphasizing that

> while the main objective in all the endeavors was to develop methods for defending ourselves against possible enemy use of biological warfare agents, it was necessary to investigate offensive possibilities in order to learn what measures could be used for defense. . . . Accordingly, the problems of offense and defense were closely interlinked in all the investigations conducted.[4]

Blurred boundaries between offense and defense continue to challenge diplomatic and government efforts to regulate bioweapons research, particularly because intimate knowledge of microbes is deemed to be the key to making weapons. These debates over dual-use research originated before the United States established a weapons program and persist in the biosciences today.

The pursuit of knowledge to wage war *against* germs simultaneously created the possibility to wage war *with* germs, building industries of biowarfare alongside expanding public health programs. Many histories of biological warfare chronicle primitive germ wars that brought a biological component to the war-making technologies of the day: Mongols catapulted plague-ridden corpses over city walls, British soldiers connived to exchange contaminated blankets during truce negotiations, and Germans infected horses and cattle with bacteria during World War I.[5] Beyond the use of weapons, the conditions of crowding and malnourishment that characterize battlefronts fostered disease transmission during war, leading to millions of deaths and generating the oft-quoted statistic that until the Second World War, disease killed more soldiers in conflict than weapons. The abundance of disease in war made it possible for militaries to strategize attacks in ways that would take advantage of disease epidemics.

As science and technology bolstered warfare in the modern era, the expanding war industry required pathogens that could be managed so as to systematically harm a selected group of people. These industries made the modern microbe into a targeted killer and consequently instilled fear in a population in ways that endemic diseases could not. Furthermore, the allied industries of war and science bracketed certain ways of knowing nature through the creation of covert weapons laboratories designed to both manufacture and contain microbes. Military-sponsored bioweapons development in the mid-twentieth century created natures that might inflict harm by the design and direction of humans while simultaneously creating spaces where the most deadly forms of nature could be safely contained for scrutiny.

The raw materials for a biological weapon, unlike plutonium or mechanical bomb components, are perceived to be readily available in nature. The persistent belief that the production of biological weapons requires high-grade knowledge of natural processes, along with free and available natural materials, places bioweapons research in the domain of the natural sciences. Bioweapons development created new categories of nature as microbiologists colluded with soldiers in creating highly technologized living agents whose intentions might be directed by humans to harm other life forms.

When the government funded this first program of remaking the microbe for military ends, it also remade human natures. The relationship of the human and nonhuman is at the heart of biological warfare. Through the work to weaponize microbes, scientists brought people's fears and desires into the material form of the germ. This work also restructured the politics of the twentieth century, creating new experiences of harm around which new human-nonhuman publics were built.[6] These new publics manufactured spaces of containment, designed to isolate humans from microbes even as they managed the microbes to do political work. In turn, these fearful politics remade practices of security and containment in terms of nature and the hybridization of germs and humans that brings about human death. The practices of the military and scientists, and military scientists, at the U.S. Army Biological Laboratories at Fort Detrick expanded what microbes could do to harm populations and laid the foundation for a politics of security rooted in modern formulations of nature.

In this chapter, I show how the alliance of war and science at Fort Detrick contributed to making the modern microbe. I consider first how the marriage of these two industries shaped the direction of the facility's work. Studying a class of microbes that could be used to wage war required remaking the spaces where microbiology was practiced, in turn shaping the type of knowledge created about the microbes themselves. The scientific work to turn microbes into weapons also shaped the modern biological laboratory.

Generating knowledge about how microbes move through air and bodies enabled the production of technologies to manage microbes in the laboratory, creating a culture of safety in which nature was presumed to be containable and human behavior always was suspect. At Fort Detrick, scientists experimented with the extremes of the human capacity to control nature. The intent to use microbes to kill underlay daily life at Fort Detrick, along with the work of caring for laboratory scientists, expanding the politics of the microbe to govern new human-animal hybrids and technoscientific natures.

MAXIMIZING THE "INHERENT POTENTIAL" OF MICROBES AS WEAPONS

The bioweapons program in the United States began in the patriotic climate of the Second World War. As one scientist remembered of his colleagues,

> We didn't believe in war. We weren't trained to be military people. We didn't believe in war of any kind, to force our will on other people, [but events like World War II] make an indelible impression on you as to the consequences of NOT having the freedom to speak, and the freedom to choose what you do. . . . In all honesty, we can't say that [World War II] was won by Fort Detrick; on the other hand, I think we would have been negligent had we not had such an extensive program."[7]

The fear of negligence prompted the U.S. government to invest millions of dollars in a bioweapons research program. The philosophical alignment of the scientific study of microbes with the work of national defense thrived in the patriotic climate during and immediately after World War II. A postwar brochure recruiting scientists to work at Fort Detrick emphasized the vital role of science in national defense:

> Research is at the base of our total defense effort. Without a constantly growing fund of knowledge, our country would soon be at a disadvantage in a cold or hot war. Research at the U.S. Army Biological Laboratories at Fort Detrick assists defense strategists in the never-ending task of keeping the United States alert and prepared by being in the vanguard of scientific development. For many Fort Detrick scientists, this contribution to our nation's defense and freedom is not shrugged off lightly. Not much is said about it, but the feeling is in the air.[8]

The research at Fort Detrick hinged first on the question of how humans could manage microbes to harm other humans and then on how one nation could harm another using germs. The possibility that microbes could be used in an act of war involved governments in microbiology. Caring for citizens required protecting them not only from epidemic disease but also from microbes used as weapons. This requirement placed the biodefense of the United States under the purview of the military. The U.S. bioweapons program had an explicit purpose: to determine what microbes might be used by Axis nations against the United States, to study how to protect citizens from these weapons, and to develop a means of retaliating in kind. These objectives aligned with those of other military research laboratories, such as the Manhattan Project, which "provide[d] scientific and technical advice in the exercise of Government responsibility for development and acquisition of new weapons."[9]

These laboratories brought scientific research into a culture distinctly different from private-sector or university settings, engaging in politics that extended beyond the production of knowledge into the manufacture of lethal weapons. Don Price describes this "union of political and scientific states" as a marriage between distinct individuals who have respect for their "quite different needs and purposes" and who "quarrel a bit."[10] However, as individuals develop into different beings through marriage, microbiological research has been reshaped through its pairing with the war industry. The "marriage" of war and science in the laboratory at Detrick created new meanings of nature (aligned with war).

In 1957, a Detrick scientist defined biological warfare as the "deliberate use of natural disease agents whose inherent potential has been exploited by scientific research and development resulting in the production of BW weapons systems."[11] By this definition, microbes inherently contain the potential to become weapons. This discourse thus dehumanizes the creation of bioweapons, locating the intent of the weapon in microbial nature rather than in weapons creators. Scientists "exploit" this harm-causing potential through the application of technologies that change the organism's state of being. This conceptualization, in which weapons systems "result" inevitably from scientific research, masks deliberate decision-making acts by scientists and implies that the work of microbiologists to create weapons merely pushes the microbe towards achieving its full potential to break through human immune defenses and do harm. The biological weapon becomes a manifestation of the microbes' capacity to destroy other life. Not only does this definition cast humans as both help and hindrance to the presumed havoc-wreaking work of microbes, but it also advances a particular knowledge of microbes as killers.

Further, the U.S. government was setting up camp and recruiting scientists in the 1940s based on a belief, speculated over in "So what?" epilogues attached to microbiology journal articles, that the killer potential of microbes could be managed to create weapons. Promoters have extolled the advantages of biological weapons in warfare and terrorism in debilitating but not killing enemies, lowering the cost of warfare, destroying life but not property, and even eliminating the environmentally destructive metals and byproducts of traditional and nuclear weapons. Microbes may be defined as killers, but they are killers that can be controlled to bring about a deliberate social outcome. This knowledge aligns with ideas of savage nature but also relies on a belief that nature is external to human social relations and subject to human control.[12] It is, in Jane Bennett's words, an "instrumentalization of nonhuman nature," whereby objects must continue to exist in subjectivity to human desires lest humans lose the moral ground for the privileging of human life over all else.[13] The biological weapon represents more than the conquest of nature, in the sense that eradication of smallpox conquered the virus, but it is a material production of a living entity that works to realize human desires, thereby upholding that ontological divide between nature and culture.

OUTCOMES OF BIOLOGICAL WEAPONS RESEARCH

Valuing microbes for their ability to harm other living organisms remade the microbe itself. The biological weapons program poured money into the study of rare microbes. Weapons scientists made deliberate decisions about what microbes they were going to study based on specific desirable characteristics. As outlined by LeRoy Fothergill in 1960, "suitable" microbes were "highly infectious," viable and stable enough "to meet minimal logistic requirements, . . . capable of being produced on a militarily significant scale," had "minimum decay rate in aerosol state," and were those to which the population was not immune.[14] Anthrax, for example, was deemed to possess the stability and infectivity to make an efficient weapon. Not only did military attention to these microbes focus broader social fears, but it also literally caused these germs to spread. Scientific study, and eventually weapons development, required the mass production of selected organisms. Scientists increased the worldwide abundance of these microbes and distributed them to research laboratories and weapons stockpiles around the world. Because of bioweapons research, anthrax exists in greater abundance and wider distribution than it would in nature.

Studying the ways that germs could be manipulated by humans to cause harm also set the stage for a new thrust in public health research. It was no longer enough to protect people from microbes they might encounter in their daily environment: now populations had to be protected against any microbe that might be introduced into their environment, particularly those whose deadly potential had been realized through weaponization. Bringing germ research to the Department of War created another agency responsible for disease control and a new reality in which public health practice alone could not protect humans from disease.

Weaponization work also created new markets for biotechnology. The United States keeps a list of "select agents," currently numbering more than sixty, identified as particularly dangerous to plants, animals, and humans and therefore likely candidates for use as a weapon. Research on these pathogens, particularly the development of antidotes and diagnostic tests, receives dedicated funding and support from government programs. As Riley Housewright, a former director of science at Fort Detrick, argued in defense of the military involvement in bioweapons, "Sometimes it's also necessary for government work because there might not be a market value that allows private enterprise to develop a cure."[15] For example, producing smallpox vaccines in the twenty-first century is not likely to be a successful business venture for a private corporation. But when the U.S. government decides to stockpile enough smallpox vaccine for all U.S. residents as a strategic act of biodefense, it creates a market value for the vaccine, and research on producing more effective vaccines continues.

The doctrine of dual use cohered beneath a complex moral umbrella that has sheltered the bioweapons program since its early years. Because the mechanisms by which microbes might be made into weapons are varied and not fully known, a defensive program begins by imagining the offensive potential of likely microbes. Once a candidate microbe has been identified, scientists might begin to study countermeasures. The seemingly limitless capability of microbes to be manipulated or mutated into weapons prompts the further study of weaponization for the sake of developing defenses. Thus defensive research also must involve work to imagine new weapons and even new microbes, work that goes against prohibitions on biological weapons.

This blurring of boundaries persisted throughout the Cold War, when the United States signed the Chemical and Biological Weapons Convention in 1972 and finally ratified the Geneva Protocol, and later, when inspectors went into Iraq to search for biological weapons labs. The Biological Weapons Convention, which prohibits the development and stockpiling of biological weapons and has been signed by more than 170 nations, permits research for defensive purposes. But where is the boundary drawn between defensive and offensive research? If research demands experimentation with offensive capabilities, the role of the researcher is continually suspect. The answer is based primarily on the scale of the research, on whether the facility is built to produce weaponized microbes on an industrial scale. Thus the research laboratory itself becomes the center of bioweapons inspections. Each flask and centrifuge and their contents must be examined and judged offensive or defensive. The laboratory is not a neutral space. Not only is its work political, but the arrangement of spaces and practices within the laboratory determine whether the microbe is an agent of harm.

MICROBES FOR PEACE: A NEW FUTURE FOR FORT DETRICK

When nearly five hundred army recruits landed at Camp Detrick to build a bioweapons laboratory, they started the program, in the words of a former science director, "from scratch," in a "ground zero effort." No weapons research "had ever been done on the scale that we're talking about."[16] Within a year of its establishment, Camp Detrick employed nearly two thousand individuals, and at its peak it had a roster of nearly 3,900, including one hundred civilians.[17] The facility studied many facets of bioweapons, including the aerosolization, mass production, storage, transportation, and delivery of live microbes. The advance teams arrived in 1943 and began to repurpose buildings and equipment, setting up experiments in airplane hangars and washing glassware in laundry tubs. With nearly $4 million available to fund new construction, work began on five laboratories and a production plant.

The conflict abroad instilled urgency in the work, and by the time the war ended, Director Merck reported to the secretary of war that "adequate defenses against a potentially dangerous method of warfare were devised, the possibility of surprise from this quarter was forestalled."[18] Merck was quick to point out that the benefits went beyond winning the war and that the U.S. bioweapons program was facilitating research on pathogens that had never before been possible. In a memo to Secretary Stimson, Merck listed eleven achievements of the program during two years of operation, including methods for detecting disease, better immunizations and understanding of immunity, facilities to propagate "tremendous" numbers of experimental animals, innovative use of photography in laboratories, and studies of plant diseases. The militarization of microbiology advanced the production of microbiology in a number of fields, yet still within the shadow of war.

When the war ended, the future of Camp Detrick was in doubt. Merck pleaded that "the metes and bounds of this type of warfare have by no means been completely measured. Work in this field, born of the necessity of war, cannot be ignored in time of peace; it must be continued on a sufficient scale to provide an adequate defense."[19] The war department decided against dismantling the camp: in fact, it purchased land in Frederick to expand operations. Between 1949 and 1955, the army replaced temporary buildings with new facilities and spent $25 million on construction. In 1956, Camp Detrick was renamed Fort Detrick and designated as the army's laboratory for biological research.[20]

The effort to weaponize pathogens created a need to control microbes as never before. Looking back on the early research at Fort Detrick on the lab's twentieth anniversary, Dr. J. J. Stubbs recalled: "There existed no precedent for the mass scale production of highly infectious agents, except perhaps by certain biological companies engaged in vaccine production. No one knew just how great the hazards would be."[21] The Detrick scientists changed not only the scale but also the attributes of microbial nature, striving to create agents that were smaller, more potent, and longer living. They altered the microbes' longevity by freeze-drying them. They studied motility and transportability to expand the reach of germ weapons. Laboratory work involved manipulating and disrupting microbial life. Both the simplest research tools, like pipettes, and the most complex, such as aerosolization systems, had the potential to release into the air a microbe that might never otherwise become airborne. Such research required a place where the germs would survive, but under containment and control.

The equipment on hand at Camp Detrick could not produce the desired level of security. This deficiency was not only a consequence of the hurried installation of the camp but a reflection of standard scientific practice. In the 1940s, microbiology was still largely done at a workbench. Scientists used simple protective equipment and practices and considered the risk of infection inconsequential in the pursuit of medical knowledge. A "tradition of personal sacrifice" prevailed among those who chose to study disease, cloaking the work with a sense of purpose, edged with risk.[22] At Detrick, however, the stakes were higher. The lab studied deadly pathogens, including "infectious disease agents which hitherto have been taboo due to the hazards involved."[23] Not only would the research bring more people into proximity with microbes, but it would focus on the microbes that harmed humans and on ways to expand that potential to harm.

Because of the increased risks, the creators of Fort Detrick aspired to make a space where it was possible to isolate microbes from the scientists' bodies—the foundation of the modern microbiological lab. In these contained spaces, scientists could observe virulence, contagion, genesis of pathogens and the effects they wrought on the living bodies of test animals without putting themselves at risk. Scientists could safely manipulate nature into the fiercest monsters they could imagine. The laboratory space also enabled the manufacturing of germs that might survive only on petri dishes, in test chambers, or, later, in DNA sequences. These new formations of life had but one environment: the laboratory itself.

INSTITUTIONAL SAFETY

During the first two years of work at Camp Detrick, sixty proven cases of infection occurred through accidental exposures to virulent agents; 159 additional exposures occurred with no proven infections. Considering the size and circumstances of the operation, Merck regarded these statistics as a testament to the safety of the laboratory. At the moment infection occurred, these infected human bodies crossed a threshold and became not only researchers but objects of research. Though accidental infection did not meet the criteria for scientific experimentation, Merck claimed that "valuable information was obtained from their treatment . . . which, but for these cases of accidental infection, could otherwise have been tested only on animals."[24] These workers all received prompt medical attention and recovered, but the accidents prompted new attention to laboratory safety and security.

Laboratory safety practices focused on creating a barrier between germs and laboratory workers. The invisibility of microbes increases the risk they pose. Consider, for example, the report of Charles Baldwin and Robert Runkle to the annual meeting of the American Association for Contamination Control in May 1967. Responding to the belief that laboratory infections are an inevitable consequence of microbiology work, they argued that accidents could be prevented through the deployment of precautionary measures:

> A new science of containment, founded on the concept of continuous agent control through the creation of intelligently designed barrier systems, has emerged. Design of these barriers is based on a rational assessment of risk; the barriers may be created in the form of solid walls, pressure differentials to control movement of air, controlled movement of personnel and materials, or inactivation of the infectious agents themselves. In the maintenance of the barrier systems one essential factor is that, at all times, the locations of the infectious agents must be known. . . . Unfortunately, such biological hazards, like radiation hazards, are usually impossible to detect by cursory examination only. Being invisible, odorless, and tasteless, they require special procedures for detection. It seems logical then, to mark the location of "biohazards," as they are commonly called, with a suitable warning sign that is readily noticed and easily recognized.[25]

Baldwin and Runkle's report shows how the barrier system was introduced and institutionalized, correlating perceived levels of risk with certain safety interventions and creating a "rational assessment of risk." They labeled this process the "science of containment."

Laboratories containing dangerous pathogens must not only be secure on the inside but also express security to those on the outside: citizens who live nearby and members of the public (see chapter 3). When the United States publicly announced its bioweapons program in 1946, the news required citizens to negotiate the meanings of the work being done at Fort Detrick. The laboratories served as tools for communicating ideas of security and containment and for helping people imagine how microbes that they could not see might be managed by technologies and the arrangement of spaces. The design of the laboratory created the belief that the most fearful agents of disease might be managed and fully controlled by humans.

When the weapons program became public, part of the disclosure involved showing people how microbes were managed inside the lab. An article in *Cosmopolitan* magazine in 1947, for example, described a tour of the "Laboratory against Death" at Fort Detrick, a "public inspection so that the taxpayers might see how they had, in a small way, repaid the government employees who gambled their lives that others may live and enjoy better health." Playing on the same spirit of self-sacrifice attributed to the scientists themselves, the author of this article attempts to ascribe ownership of the laboratory to the taxpayers, describing the sacrifice of laboratory workers as a debt for which every citizen must pay. The National Institutes of Health director, Rolla E. Dyer, is the hero of the article, acclaimed for the dogged pursuit of maximum safety. The article also quotes W.E. Reynolds, commissioner of public buildings, who said the facility did not expect to eliminate "every last laboratory accident among scientists working day in and day out with dangerous cultures" but that the new laboratories would "eliminate some of the danger to many innocent bystanders."

Williams's article about the new bioweapons lab seems to have received a lot of attention in its day. It was reprinted in *Reader's Digest*, and several community newspapers reported that the article was read at various public meetings. After the public tours ended, the doors to the laboratory were closed: "Each cell block now has been closed forever to all except qualified workers, their guinea pigs and their ultra-microscopic enemies. The fight has been resumed on the new battlefields."[26] The physical barricading of the space is necessitated by the qualities of the microbes, but it also enhances secrecy. The process of containment defined the microbe as an entity rendered harmless by separation from humans but made deadly through mere contact.

Building the modern germ laboratory at Fort Detrick also built the belief that microbes could be managed through technology and human behavior. Containment was key. As one laboratory safety director explained in a scientific publication, "Microbiological environmental control, the objective of good design, involves any technique, equipment or building feature or combination of these that confines microorganisms within a specific environment."[27] Containment practices created distinct "environments" defined by separation from humans, even when in close proximity. These environments were separated by barriers that inhibited connections between living organisms. However, the barriers did not inhibit the work of the microbes. Instead they became part of the identity of the microbe, a way of defining how germs moved through environments and connected with living matter to cause infection and pursue their own survival. Confining microbes meant producing them as objects that could be managed. The microbes, however, were agents in creating these laboratories. They participated in a politics of the laboratory that created new natures and new systems of governance and security.

The remainder of this chapter looks closely at the particular characteristics of the modern microbiology laboratory that were made at Fort Detrick. I consider specific technologies as they relate to the regulation of airflow, the life and death of laboratory animals, and human behaviors with accompanying social controls. The design of these laboratories made it possible to believe that technologies could manage the human-microbe relationship in pursuit of cultural objectives. In this space, the crossing of boundaries was taboo, and hybridization was something to be managed and feared. Scientific knowledge made in these spaces created a politics of life and death whereby the inability to control nature violated boundaries and spelled illness and death. "Microbiological environmental control" created a focus on separation and on the construction of an environment specific to an organism and separate from the space occupied by humans. This work hinged on an environmentally delineated dualism between nature and culture and a potentially deadly exchange of existence between germs and humans. Movement across barriers was suspect and became the target of a technological fix. A closer look at these technologies shows how they also constructed microbes as harmful agents that move through time and space with the objective of infecting and parasitizing human life. All microbial activity is suspect and is understood first in terms of how it might be managed.

AIR: ANIMATING MICROBES

Some of the fear surrounding microbes comes from the perception that deadly pathogens are floating invisibly in the air. This belief recalls understandings of disease that predated the germ theory, such as the view of miasma, or "bad air," that carried particles of rotting matter. Knowing that infection can occur through inhalation transforms the life-sustaining act of breathing into a risky behavior. The work of the twentieth-century bioweapons program furthered this fear of air. Because aerial dispersion has long been seen as a suitable method for dispersing a pathogen within a population—particularly an organism such as anthrax, which is not communicable from human to human through contact—one category of biological "weapon" fills the air with microbes that the population cannot avoid inhaling.

A branch of research at Fort Detrick focused on how to make microbes that were more stable in the air and would survive field dissemination, increasing opportunities for infection. Scientists experimented with microbes to study the relationship between the size of particles inhaled into the lungs and the initiation of infection. Fothergill summarizes this relationship: "Very small particles, in a size range of one to five microns in diameter, are capable of passing these impinging barriers [the turbinates of the nose and cilia of the trachea] and entering the alveolar bed of the lungs; an area highly susceptible to infection."[28] For example, particles of anthrax one micron in diameter could cause infection at a dose seventeen times smaller than particles twelve microns in diameter.[29] Studies of other agents showed even more dramatic increases in infectivity with smaller particles. Experiments also showed that smaller particles remained suspended in the air for longer periods, traveling farther before losing infectivity, and that "an aerosol of such small particles will diffuse through structures in much the same manner as a gas, thus giving it a remarkable property for target searching."[30] Weaponizing an infectious agent, then, required producing microbes that were stable in small spores. The fascination with anthrax as a potential bioweapon, for example, comes from its very small spores, which travel through the air and deep into the lungs to produce infection.

In 1947, construction began on an aerobiology chamber at Fort Detrick, an airtight sphere of stainless steel nearly four stories high, which still stands today. The round chamber, dubbed the Eight Ball, has walls 1.25 inches thick and a volume of one million liters. Its equator is lined with biological safety cabinets and cubicles where samples could be collected and animals could be exposed to substances released in the chamber. Humans could breathe the air inside through masks.[31]

The Eight Ball enabled scientists to bring the air itself into the assessment of how microbes move. Devices at the base of the globe projected experimental substances into the air, and their movements through that space were measured by sampling around the wall of the sphere. The application to place the test sphere on the National Register of Historic Places after it was decommissioned in 1970 describes the site as formative in science practice: "In essence, the staff of the old Biological Laboratories weaned aerobiology from a qualitative into a quantitative science, and the Test Sphere not only was instrumental in this accomplishment but also stands as a tribute to the scientific staff that did it."[32] This maturing of science into a quantitative practice subjected air to scientific scrutiny. It could be measured, controlled, talked about, and written about in ways that influenced scientific inquiry. Moreover, the air itself became an element of the biological weapon, a means of dissemination.

Tests in the Eight Ball could not replicate environmental conditions such as sunlight and atmospheric turbulence. In 1946, Camp Detrick annexed nearly four hundred acres of land in Frederick to create an open-air space for testing. The land, known as Area B, was mapped in seven concentric circles around a central point from which aerosols could be released. Sensing devices placed around the grid detected the

FIGURE 7. Fort Detrick technicians operate the "Eight Ball" aerosolized test sphere, studying how people, animals, and objects take in microbes dispersed through the air. Image courtesy of the Department of the Army.

aerosols, turning microbial movement into data sets mapped onto the landscape. Like the larger-scale proving grounds in Mississippi and Utah for testing biological weapons, studies at Area B produced knowledge about how microbes move through the air.

Perhaps the most publicly controversial work of the bioweapons program consisted of field trials designed to track the movement of biological agents by mass-producing them and then releasing them into the inhabited environment. Not all of these tests took place at the proving grounds. One test, for example, sprayed nonpathogenic spores of *Bacillus subtilis,* var. *niger,* an organism used to imitate anthrax in bioweapons research, from hoses on the deck of a boat floating two miles off the coast of the United States. A network of sampling stations set up in the homes of government employees and government office buildings detected spores twenty-three miles from the source.[33] A similar test used fluorescent particles of zinc cadmium sulfide released from a ship 10 miles offshore along a 156-mile course. Tracing where the particles landed on shore showed that the aerosol dispersed over an area of 34,000 square miles and traveled about 450 miles from the source.[34] In these types of experiments, the environmental circulation of air was scrutinized along with the mechanisms used to transport and release biological agents. Looking for traces of the zinc aerosol using ultraviolet light made visible the reach of a

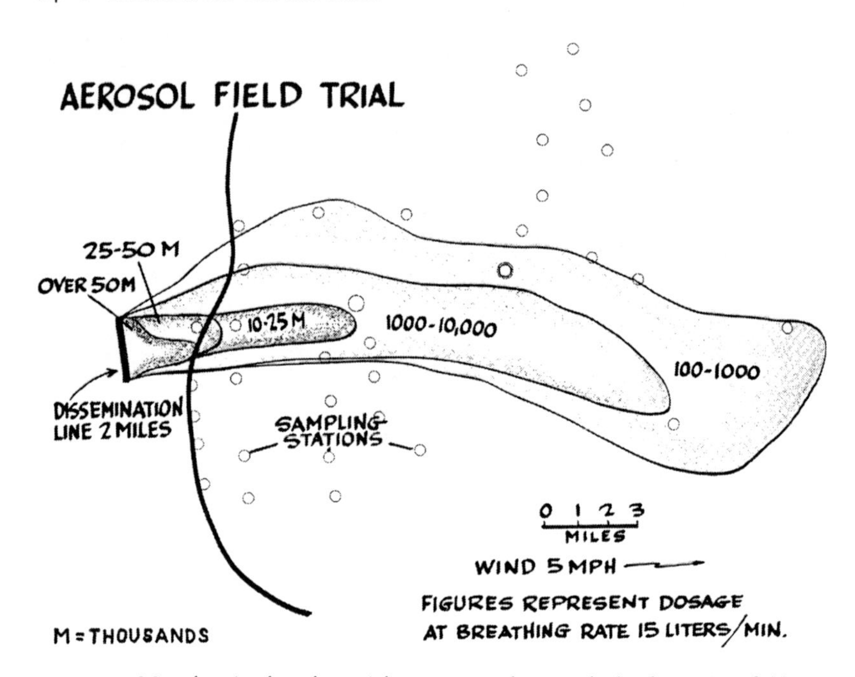

FIGURE 8. Map showing how bacterial spores spread across the landscape in a field test and the numbers of spores that might be inhaled by humans. Leroy Fothergill, "Biological Agents in Warfare and Defence," *New Scientist*, November 30, 1961, 546.

biological cloud, while the *B. subtilis* trials demonstrated that living organisms could travel through the air into homes and offices. The experiments animated unseen life forms and reified the potential for distant pathogens to move into people's private spaces. Though controversy stirred in later years about the government's decision to release supposedly innocuous microbes on citizens without their consent, these field trials marked people's homes and communities as potential targets of airborne microbes.

The final reports contended that these field trials were rudimentary, using delivery devices that were pieced together from commercially available products, and that scientists (or national enemies) could easily have used a greater concentration of biological material. Fothergill concluded that "it requires little imagination to conceive of . . . the design of efficient military equipment to accomplish this. . . . One's imagination can run wild."[35] The trials instilled fear of microbes moving unseen in the air, with the type of contagion experienced through one-on-one human contacts now expanded to communities or even the nation-state. These experiments represented microbes as an invisible cloud, growing ever larger and spreading farther through the work of air. This was a potent metaphor for imagining the microscopic world.

Because the cloud could be seen only with scientific tools, it privileged the knowledge of experts for identifying the threat. One researcher reminded readers of the dangers of trying to detect the cloud: "Biological clouds have no characteristics detectable by the senses. They are invisible, odorless and tasteless—in contrast to certain gas clouds. Even if they possessed an odor, the odor-detecting sniff might result in a sufficient dose to produce an infection."[36] The biological weapon, like its nuclear counterpart, builds fear because the moment of contact with the body may pass unnoticed, and its presence becomes visible only as the human body deteriorates. As with nuclear fallout, the infective cloud carries unseen agents of harm many miles beyond an initial site of impact. During a time when scientists were also estimating the reach of nuclear weapons, the graphic representations created by bioweapons scientists helped create the cultural imaginary of the weapons of the Cold War, broadening the message of fear to one of large and invisible clouds of death.

Part of the fear of biological weapons emerges from the human reliance on air, even when that air may bear pathogens that cause harm. Managing the risk of microbes demands the control of air. While the ability to control airflow in civic spaces has yet to be realized in the same way that water flow has been technologically managed in modern societies, it has been achieved in the microbiology laboratory. Tight controls of laboratory air are used to manage microbial movement and reduce the risk of infection among laboratory workers. In the laboratory, technologies and the behavioral practices allow bodies to be near pathogens without producing opportunities for infection. The laboratories at Detrick, where scientists handled highly infectious agents on a daily basis without contracting a disease, demonstrated the modern ideal that disease can be contained through spatial configuration.

TECHNIQUES AND TECHNOLOGIES TO CONTAIN MICROBES IN THE AIR

The scientific work that aerosolized the microbe also created the basis for laboratory security, which is primarily produced by regulating the flow of air. As the 1947 *Cosmopolitan* article explained, "The air moves from the central hall into the decontamination lock when the door is opened. It also travels from the laboratory corridor into the workrooms. As a matter of fact, it moves away from the lab worker himself at all times except when he's on his way out."[37] Not only does this description create a powerful image of germs being sucked away from human bodies, but it further establishes that microbes moving through the air create the danger of transmission.

Notably, in the laboratory it is primarily the work of scientists that makes germs airborne and therefore risky. In the mid-1950s, Briggs Phillips and Morton Reitman conducted a series of experiments designed to identify how microbes became airborne during laboratory work.[38] They sampled the air around the experiment while the scientist was working, calculating how many microbes were released into the air by different activities such as pipetting, stirring, grinding, or mixing. Using new photographic techniques, Johansson and Ferris showed that blowing out the last drop of a cultured solution onto a petri dish formed bubbles at the tip of the pipette and aerosolized the microbes.[39] The risk of infection that resulted from this scientific practice was mitigated by both behavior and technology. For example, laboratory protocols suggested not forcing that last drop from the pipette. Containment hoods were built to separate workers' faces from the microbial solutions with which scientists worked.

The design and creation of laboratory safety cabinets is often cited as one of the significant contributions of the bioweapons program at Fort Detrick. When work began there, microbiology was a workbench science, borrowing equipment from other fields such as chemistry. Hubert Kaempf, a soldier assigned to the sheet-metal department, is now credited with creating the first class III safety cabinet, working up designs to materialize ideas imagined by the microbiologists.[40] The image of a rectangular glass box with two rubber gloves attached to the front, where people can insert their hands to work with the contents, is now emblematic of the biolab. The biosafety cabinet contains microbes within its chambers—sometimes a small chamber where one set of hands works, and sometimes a series of chambers, perhaps refrigerated or heated for different stages of an experiment, connected to each other but sealed from the outside. Every material removed from the biosafety cabinet to the main laboratory space is sterilized, whether through chemical treatment, ultraviolet radiation, or heating in an autoclave. No life form leaves the microenvironment. A steady flow of air into the safety cabinet contains the microbe, while outflow air is sterilized to remove life. The system is the reverse of a gnotobiotic cabinet, where the air flows outward to prevent the contamination of the environment inside.

The popularization of the air-driven biological safety cabinet during the bioweapons program correlates with the work that made airborne

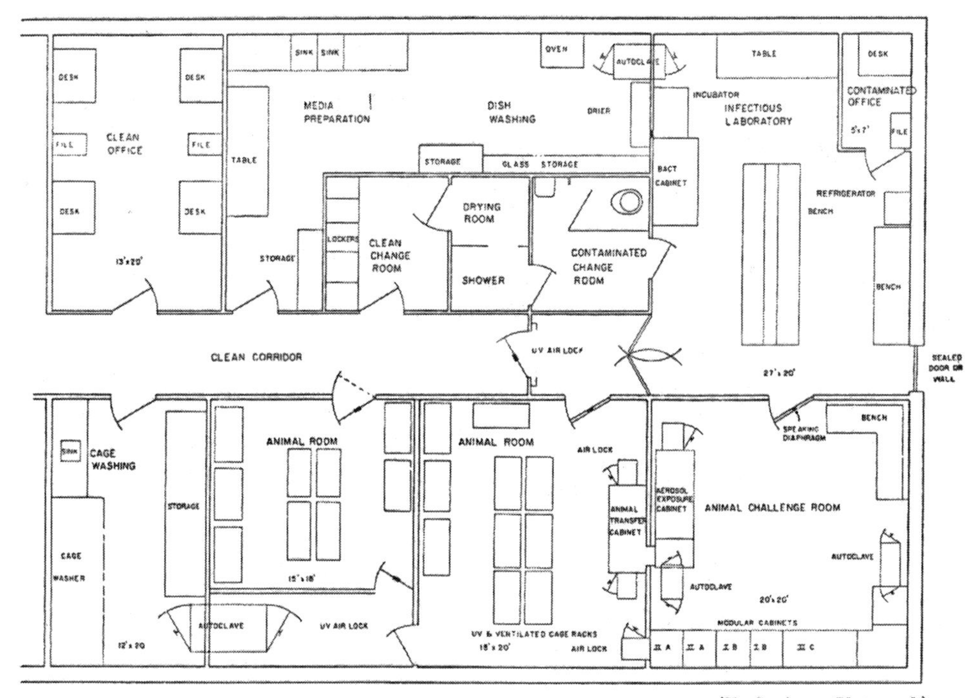

(U. S. Army Photograph)

FIGURE 9. Floor plan for a biosafety laboratory. Arnold G. Wedum, Everett Hanel, G. Briggs Phillips, and Orrin T. Miller, "Laboratory Design for Study of Infectious Disease," *American Journal of Public Health* 46 (1956): 1111.

germs into long-distance weapons to begin with. A 1956 article describing laboratory safety practices identified the bacteriological safety cabinets as the most vital safety component of the biolab, citing as evidence a case in which a building's air sterilization system was rendered "accidentally inoperative," but no laboratory workers or "passers-by" were infected because the cabinet air was being sterilized first.[41] The inadvertent failure of one component of the system created an opportunity to demonstrate the effectiveness of technologies that controlled the worker's immediate environment. Indeed, redundancy is the foundation of systems to contain microbes in the laboratory: much of the air leaving a high-security biolab is filtered three times: through the safety cabinet, as it exits the workroom, and again as it leaves the building.

At the scale of the workroom, the regulation of airflow mimics that of the safety cabinet. The room is kept at a negative pressure, drawing a steady stream of air into the room. Like anything leaving the room, outgoing air must be sterilized, circulated through a series of filters to capture ever-smaller particles. The need to control air movement within a laboratory building led to the frequent publication of floor plans for biolabs in the microbiology journals of the mid-twentieth century.[42] The definitive two-volume manual *Design Criteria for Microbiological Facilities at Fort Detrick, Maryland* spelled out all aspects of laboratory design in terms first for management personnel and planners, and then, in great detail, for the architects and engineers.[43] The second volume gave specifications for everything from the paint, pipes, and plugs to the ventilation system and instrumentation within the laboratory. Even the materiality of the building comes under scrutiny for safety, for a wall that appears solid may be porous to microbes and air and therefore must be sealed. This manual, which was distributed to laboratories around the country, contributed to a new understanding of biosecurity as the work of constructing spaces to manage the movement of life forms inside, entwining the physical architecture of a laboratory with the management of risk. The thoughtful use of markers that characterized the biosecurity of this space—right down to the smooth walls coated with high-gloss paint to facilitate the capture of escaped insects—was encouraged in the manual under a broader discussion of design philosophy. A series of ideas, actions, and material products had come to represent biosecurity in the cultural imaginary. Blueprints made biosecurity visible and intelligible in the design plans of the architect, creating a system anyone could follow to isolate environments and manage microbial natures. The iconic images of hooded scientists stepping from the laboratory into a sealed shower room, where they are doused in disinfectant, provide a stronger visual image of biosecurity than air moving through ultraviolet lights and filters, but both depictions demonstrate the work of isolating clean spaces from contaminated spaces within the laboratory.

The complexities of biosecurity measures increase the costs and the consequences of disruption. These systems have to operate around the clock, but like any technology, they are susceptible to breakdowns and malfunctions. Because all the air in the building is filtered, the filters must catch particles ranging from lint, hair, and fur to microscopic germs. Clogged filters must be replaced. But how can the building's airflow be maintained while the filter is changed? Duplication and backup systems are required, which increase the cost. If, as some have argued, these elaborate systems provide little more protection than the bacteriological safety cabinets, what is the rationale for their use? Reitman and Wedum argued in 1956 that these systems may be "necessary for public relations or legal reasons," suggesting that the public's perception of security may have value in the production of the laboratory space.[44] I explore this idea further in chapter 3.

While numerous articles promoted the new biosafety equipment created and tested at Fort Detrick, many also hedged the promise contained in these technologies with a reminder that biosecurity also depends on human behavior. Following their claim that the safety cabinet is "the most important single piece of equipment in preventing laboratory infections," Wedum and colleagues say that "of course, it cannot substitute for good training and good laboratory protocols."[45] Humans can violate the physical security created through laboratory technologies, though the proudest systems claim to overcome human fallibility or ill intent.

ANIMALS: CREATING RISKY HYBRIDITY

Along with managing the movement of air and matter, laboratory spaces also manage the movement of animals. Animals kept in laboratories and infected with pathogens require handling to sustain their life and to manage their bodies as experimental sites. After the initial act of infecting an animal with germs, whether via air or another mechanism, it must be fed, its enclosure cleaned, waste disposed of, interactions with other living organisms regulated, and many aspects of its physiology measured to generate data. These activities create multiple sites of risk from contact between human hands and animal-microbe hybrid bodies, creating potential for transmission of disease through air, blood, and flesh. In a 1952 survey of 1,087 laboratory infections, animal bites and scratches accounted for 12 percent of infections.[46]

Because contact between human and nonhuman species is vital in animal laboratories, the risk of handling and cross-contamination must be minimized. Wedum's laboratory floor plans for experiments involving animals include spaces for "normal animal holding" isolated from the "animal challenge room," which in turn is connected to a clean "cage washing room" by an autoclave that destroys microbial life. The authors recommend that small animals that are not in experiments involving airborne microbes be kept in ultraviolet cage racks, where UV lamps create a "radiation barrier" that kills airborne microorganisms as they leave the cage.[47] Handlers must wear goggles to shield their eyes from the radiation. One design provides access to the animal room through a door with an ultraviolet barrier, subjecting the human body to a similar dose of ultraviolet exposure. In studies involving infection through the respiratory system, scientists are cautioned to remember that airborne organisms settle into animal fur and that animals may slough off pathogens for six or more days after exposure. For these experiments, the ventilation of cages is controlled.[48]

Hybrid test subjects, part disease and part animal, create infected waste in the form of fluids, feces, and ultimately carcasses. Living microbes in the waste from animal bodies must be rendered harmless before the waste can be treated through traditional waste-management systems and removed from the laboratory. Similarly, when an infected animal dies, the carcass must be treated to kill the microbes living in it.

Laboratory designers also are concerned with interactions between animals. Not only must uninfected animals be managed to prevent unplanned infection, but the many hybrids must be kept in isolation from each other to control variables of infection and contagion. Furthermore, animals may carry latent infections, either brought to the laboratory in their bodies or acquired in the laboratory space. Although these may not be manifest in the animal's flesh, they may shape the outcomes of experiments or spread among animals through contact in the laboratory.

The size of animal bodies shapes decisions about laboratory design. Designers caution that animal-based research requires significantly more secured space, increasing the cost of laboratories significantly—a cost proportional to the size of the species being handled. Mice require different spaces from goats or primates. In Wedum and colleagues' floor plans for animal labs, half the laboratory space is dedicated to animal handling. Because of the high costs of biosecure space, the infected animal body is created at significant expense. The desire to observe disease on and in animal flesh creates in the laboratory new forms of risk that are mitigated through dedicated spaces and handling practices.

HUMAN BIOLOGY AND BEHAVIOR AS A SITE OF VULNERABILITY

The objective of biosecurity is to prevent the infection of humans by the pathogens being studied. As one "state of the art report" on "microbiological contamination control" described the risk, "Because man is an extremely prolific source of microorganisms, and because he can be extremely susceptible to those microbes that are pathogenic for him, inclusion of man in any system usually signals the weakest point in the contamination control effort."[49] Responding to the proposed need to "deal simultaneously with a series of interlocking factors," Jemski and Phillips proposed a "'systems approach' to safety which consists of man, his environment, and accident agents, [and where] the ecology of the situation is examined and manipulated so as to put man in a favorable survival position."[50] They put forth a list of approaches to mitigating laboratory risk, beginning with the vaccination of laboratory workers. Immunity created through vaccination mitigates risk at the site of the human body, bringing it under the management of the laboratory.

Notably, however, immunity is measured by the human body's ability to resist disease through antibodies, which are generated by exposure to pathogens. Bodily intake of a sufficient number of germs, whether through contact in the environment or deliberately drawn in through a controlled dose of vaccine, prompts the human body to produce the means to resist infection. However, a vaccine designed to protect the general population may be inadequate to protect the body against germs encountered in the laboratory. As one National Institutes of Health biosafety brochure reminded workers, "Under laboratory conditions, infectious agents are usually handled in high concentrations, and infections may be acquired by uncommon routes."[51] Inhaling anthrax, for example, is an uncommon way to contract this soil-dwelling organism. Laboratory workers also may be exposed to microorganisms in higher concentrations than a human might normally expect to encounter. Immunity is not an absolute state but exists in degrees. The unique environment of a laboratory, created to promote to the generation of new microbial life, also creates unique pressures on the human immune system.

Mandatory immunization was strongly advocated by early developers of laboratory safety protocols, who often presented it as the first and last barrier against infection. However, requiring workers to change their physiology through a process declared safe, but which also creates risk, raises questions about the reach of the laboratory into its community and into human bodies. Wedum and Phillips present a list of twenty-seven "questions on policy" in their 1964 article "Criteria for Design of a Microbiology Research Laboratory," prompting designers of laboratories to calculate a risk and consider how a practice or space within a laboratory could mitigate it. Their format presents each item as a possible action connected to specific outcomes. Under the question of personnel policy, the article asks whether employees must "accept vaccination with commercial standard vaccines and with experimental vaccines when, in the opinion of the laboratory director, administration of these would decrease the chance of clinically apparent illness? A yes answer will reduce the need for mechanical protection." By mandating an intervention using workers' bodies, then, laboratory managers might reduce the cost of laboratory design. A second question asks "What level of occupational infection is acceptable to management?," reminding managers that not all laboratory infections will manifest as disease and that the "minor discomfort" of vaccinations may also affect labor output.[52] Notably, one of the criteria for classifying a modern laboratory at the highest security level, biosafety level 4, is the study of agents for which no immunological interventions are known. This renders moot the question of vaccination. Still, instituting mandatory vaccination as a means of laboratory safety illuminates the reach of the biosecurity system into individuals' lives, often at the discretion of decision makers weighing value and risks. This series of design questions also shows how security is constructed in layers through the control of space, elements, and humans.

Another crux for laboratory safety policies lies in the regulation of human behavior and the ability of laboratory managers to control employees' actions. One article on lab safety from 1959 reiterates the mantra that "each person is responsible for preventing accidents during the course of his individual actions."[53] Mitigating risk requires managing individuals' behavior while working in the laboratory—stopping them, for example, from extracting that last drop from a pipette. Wedum and Phillips argue that laboratory safety practices are easier to implement if they do not require people to work differently but either happen before or after the work begins or are passive, requiring no deliberate action. They also observe that if the laboratory employs large numbers of "nonprofessional personnel," there is a greater need for safety barriers within the space. Smaller labs with people who have better "judgment based on education and experience" require fewer mechanical interventions. The design of the space, then, should create security precautions that cannot readily be undone by human behavior, even while human behavior is the linchpin of the security system.[54]

A laboratory designed around the scientific calculation of human behavior takes shape around the categories that organize all collective action. Gender, for example, defines part of the laboratory space: a laboratory that employs more men than women may design a space with a larger changing room for men, if they include a space for women at all, creating the perception that the laboratory space cannot accommodate women scientists. Reitman and Wedum propose that laboratory workers wear "suitable laboratory clothing," specifying that men should wear an operating gown that ties in the back—a clear association with medical professionals—while women should wear a smock.[55]

The social connection between gender and behavior is also revealed in a safety system predicated on the management of human actions. A biohazard containment course at Fort Detrick in the late 1970s discussed the need to police behavior to create security and offered several examples of potential problems. Trainees are cautioned about "the Lone Wolf," who may defy protocols and peer pressure, as well as "females as authority figures." The training handouts suggest the following role for the woman scientist: "Due to sexual stereotyping within our culture, there are certain behaviors which a woman who is in a position of authority should try to avoid. . . . She must behave as a technician or a scientist in the laboratory, which is an asexual role. That is, she should never attempt to manipulate males seductively. If she does she will relegate herself to the wife-mistress role, which is not one which engenders respect in the working environment. Manipulation should always be through expertise on the job, and hard work." While the woman scientist is still assumed to want to manipulate her peers, she is expected to strip herself of gendered behavior as she dons her scientist's smock in order to access a respect that is not afforded women more generally in the population. Notably, this advice is built on the assumption that expressions of gender or sexuality in the workplace—specifically, the loss of authority by a scientist who behaves like a woman in the workplace—creates risk and threatens the security of the space. The patriarchal ordering of the laboratory is rationalized by the high value placed on security and the perception that biosecurity can be created only through the precise and authoritative control of human behavior.

All forms of difference are suspect in a system focused on preventing disruption. A unit in the same course, titled "Minority Problems," listed strategies for collecting information on what minority employees are doing in the laboratory and then offered the advice to "keep them busy, separated, and positively reinforced."[56] Deviance, including deviance by race or gender, creates risk and is therefore subject to management for the sake of security, turning the laboratory where microbes are handled into a small state of exception where people are subject to unjust rule for the sake of securing the population.

Numerous laboratory safety manuals published at Fort Detrick and elsewhere lay out guidance for desirable worker behaviors. The culture of self-sacrifice repeatedly identified among early microbiologists led to reports of workers who continued to believe that rules and regulations were unnecessary and that risks were worth taking, as well as suspicions that people were not reporting accidents in the laboratory. Wedum also reported that the average laboratory worker "was primarily concerned with what he felt was personal involvement in safety matters, such as his own safety, avoidance of guilt in the matter of infecting fellow employees, [and] annoyances involved in safety procedures."[57] Employees could be motivated to protect their own safety and to some extent the safety of others if it wasn't too much trouble, but Wedum concluded that workers were not interested in changing their behaviors to increase biosecurity. Despite the high cost, managing spaces and equipment to create security and override deviant behavior was deemed to be easier than managing human behaviors, though safety protocols continued to be written, published, and distributed among laboratory workers.

OUTCOMES: CONTAINMENT AND SECURITY

The bioweapons program at Fort Detrick operated until 1969, when President Richard M. Nixon formally ended offensive biological weapons research in the United States, renouncing bioweapons because of their unpredictable nature and inefficacy as a tool of war.[58] As Detrick scientists closed their laboratories and tried to redirect their research toward a more peaceful purpose, the U.S. government debated what to do with the laboratory buildings themselves. Operating the facilities built to contain the world's deadliest microbes was expensive, and the buildings were difficult to repurpose. Indeed, the extent to which the alliance between war and microbiology had shaped the laboratories at Fort Detrick became abundantly clear when the government tried to reuse the space for less secure and secretive science programs. The bioweapons facility was housed on a military base, then under the direction of the U.S. Army Medical Research Institute of Infectious Diseases (USAMRIID). Nonmilitary scientists were unwilling to be associated with the biological weapons factory. In 1971, by presidential proclamation, the Frederick Cancer Research Center of the National Cancer Institute took over the space, a decision that Nixon declared would fulfill his pledge to turn swords into ploughshares.

Though the building was repurposed, the body of knowledge created at Fort Detrick originated ideas of containment and human-created risk that shape laboratories today. The bioweapons program demonstrated that containment was possible through diligent work to construct environments and manage human behavior. Weaponization created both the possibility and the requirement that microbes could be contained. Through new technologies, the laboratory became a landscape where microbes might be managed according to human objectives. Humans came to be seen as the weak point in biosecurity, for the barrier effects of the technologies could be diminished by human behaviors, particularly carelessness or neglect. Human and animal bodies create risk in the biolab, and if infected they enter a system of containment that they cannot escape until declared clean, even to the extent of cleansing carcasses of microbes before crossing thresholds that divide the unclean from the clean. In turn, this belief in containment, materially expressed in the modern biolab, shaped governance throughout the twentieth century. The promise that the nation could protect its citizens by containing threats grew to encompass cross-boundary terrorism and unknown microbial menaces.

The biolab also draws attention to the vitality of circulation, for not only is air circulation necessary for human life, but the microbes and animals being studied in laboratory spaces demand an environment that sufficiently mimics nature to keep them alive.[59] The work done at Fort Detrick to build secure biolaboratories was about containment, but it was also about circulation and movement. Microbes moved through Fort Detrick in the social work of national security, and these circulations had to continue for the lab to produce results. Care did not simply mean containment: it meant the continuance of work in an environment where citizens could expect physical barriers to protect them against microbes.

The controlled circumstances of the laboratory made it possible to imagine a broader landscape in which germs could be expertly controlled. The work at Fort Detrick created both the possibility that microbes could be managed by humans to inflict as-yet-unimagined harm and the tangible spaces for testing and exploring the technologies for doing so. The work done there also established the responsibility of scientists to communicate with the public about the security of the laboratories, an articulation of security that established the government's caring role of watching over citizens' health. The study of deadly pathogens was increasingly seen as important work that would protect the nation's future—a role affirmed by the laboratory's military origins and the pairing of microbiology and weapons research. It was a military obligation to study the microbes that could harm Americans and to enhance the capability of these microbes to secure the borders, whether through the production of a biological superweapon or an undefeatable vaccination.

The final days of the bioweapons program were marked by the destruction of the weapons developed at Fort Detrick. Scientists incinerated their experimental samples, testing for surviving organisms before mixing the ashes into the dirt at Area B. The dust was sterile, rendered devoid of life by one of the many technologies that secured the laboratory. Years later, environmentalists would question the effectiveness of the sterilization process, and the work to destroy the stockpile at the Dugway Proving Ground dragged on for years in debates over how to most effectively disarm a weapon designed to be indestructible. Although the pathogens were, at least in theory, destroyed, the knowledge of biological warfare and the technologies like the safety cabinet would survive the dismantling of the weapons program.

The end of a state-sponsored program at Fort Detrick also created a social context for a new brand of biological weapons scientist: the iconic rogue scientist concocting infectious bombs in the basement with microbial specimens ordered off the Internet and a knowledge of college-level biology. The modern weaponized microbe might have been autoclaved, destroyed in a distant desert, or preserved in a freezer, but it could not shed its militant associations. The belief that microbial nature is freely accessible (even to the terrorist) masks the system through which that nature is produced. The germ we imagine brewing in the makeshift biolab would not exist without the high-security laboratory space designed to make the most dangerous germs in the world.

3

The Wild Microbiological West

Fighting Ticks and Weighing Risks

On February 14, 2002, Mary Wulff, a retired police officer who had moved to Hamilton, Montana, from Los Angeles, read an announcement in the *Ravalli County Republic* about a public meeting to be held regarding changes at Rocky Mountain Laboratories (RML). At the meeting, she heard lab officials announce to the two dozen attending community members that a $66 million grant from the federal government would be used to build an Integrated Research Facility (IRF), a swank new structure designed to study the most dangerous pathogens on the planet in high-security biosafety level 4 (BSL4) laboratories. Wulff recalls looking around the room to assess whether anyone else was as startled and alarmed to hear this as she was. "Until that day, I had never really thought about the lab before. I had never thought that this would happen in my town."[1] She sat through the meeting in a daze, unsure what to say or how to respond. Afterwards, she went home and started calling people. Within weeks she had established a nonprofit group, the Coalition for a Safe Lab, and was networking with other groups to probe what was happening in the red-brick buildings west of Fourth Street.

Over the next four years, Wulff became fluent in topics ranging from emerging infectious diseases to biocontainment, as well as the finer points of environmental impact statements and the Freedom of Information Act. The grassroots movement swelled as local environmental groups joined the Coalition for a Safe Lab in demanding transparency

FIGURE 10. Aerial view of Hamilton, Montana, showing the Bitterroot River and the Rocky Mountain Laboratories on the east side of the river. Photo: National Institute of Allergy and Infectious Disease, National Institutes of Health.

in all stages of the planning process. Although community activism did not stop the construction of the BSL4 expansion, both laboratory officials and grassroots protesters claim that the movement played an important role in shaping the first BSL4 laboratory to be built since the terrorist attacks of September 11, 2001. Not only did the physical plan for the laboratory change to address residents' security concerns, but the protest also set a precedent for community involvement in evaluating the risks of cultivating deadly microbes in proximity to healthy citizens. The controversy affirmed that RML was not just part of a national science community but also an integral part of a local community of human and nonhuman agents, whose future was invested in complex ways in the materiality of the laboratory and the work done there. Because of the protest, RML officials reexamined the responsibilities of the laboratory to the community, while Hamilton residents negotiated the economic and cultural effects of the science industry in their town. These discussions revitalized debates that have simmered throughout the lab's hundred-year history: community resistance, local and national public health pressures, and trends in disease science have continually remade Rocky Mountain Laboratories. The lab became a place where both scientists and citizens claimed a role in the identification of risk.

The contestation of the IRF in Hamilton shows how the notion of biosecurity as a set of technospatial practices that contain microbes for scientific study has been reworked in the twenty-first century to encompass the elimination of vulnerability to all biological threats. The ideal of biosecurity that originated on a small scale in the production of the biolab emerged from and reinforced a broader social belief that technology can contain and control natures. These beliefs are mutually constituted and create a powerful connection between security and technology, which forms the basis for preparedness work in the modern security state. Moreover, the political ecology of the Bitterroot Valley in Montana has repeatedly brought human and nonhuman subjects into conflict over space. In Hamilton, *Rickettsia rickettsii* (the bacterium that causes Rocky Mountain spotted fever), the ticks that carry it, the Bitterroot River, cows, farmers, scientists, and activists move through a contested landscape in ways that have required humans to continually rethink how to live with germs. In a place known for environmental risk from its earliest homesteading period, citizens established institutions and mechanisms for assessing hazards. In the IRF controversy, these mechanisms were used to hold government accountable for mitigating the new types of risk that microbes might pose to human life. The federal project to prepare the nation for terrorist attacks after 9/11 created a new paradigm of risk in Hamilton. By contrasting the twenty-first-century protests in Hamilton with earlier conflicts, I show how a particular sensibility emerged alongside the rising "war on terror" and new microbial formations.

Not only were the Rocky Mountain Laboratories physically and ideologically remade to meet the new threats of bioterrorism, but citizens of Hamilton also created new categories of citizenship predicated on changing natures and notions of risk. Numerous theorizations of biological citizenship might be summed up as expressions of people's ability to claim a right to life and protection of that life.[2] Nikolas Rose and Carlos Novas have argued that genomic biology has created new manifestations of citizenship along with "new spaces of public dispute . . . new objects of contestation . . . novel forums for political debate, new questions for democracy and new styles of activism," which are informed by unique cultural histories, forms of governance, and activist traditions.[3] Whereas prior conceptualizations of biological citizenship emerged from the scrutiny of unhealthy bodies damaged by environmental catastrophe or carrying genetic disease, the new biology demands consideration of biocitizenship as the acts of healthy subjects who come together to dispute state actions and advocate for biosecurity. The history of RML shows that people have long understood biological risk in terms of what might happen to their bodies and have advocated publicly for mitigation of that risk through the court system, public procedures, and personal interactions.

These actions bring material change to the spaces where science is done. The events in Hamilton also exemplify how grassroots organizing and institutional change shape biocitizenship debates. At a time when the pastoral role of the state in health care is diminishing and responsibility for collective health is being transferred to individuals and corporations, citizens advocate for a particular form of government care centered on security. Thus, local actions constitute national biosecurity.[4]

This chapter begins by exploring the origins of the Rocky Mountain Laboratories and the geographies of disease that entrenched microbiology in the social structures of the community. It then considers more broadly how today's Hamilton citizens assess risk. Guided by Paul Rabinow's theorization of biosociality, I examine the groupings of activists, scientists, government, and community that formed during the protest against the RML expansion in order to understand how the new natures of this age of bioterror created a crisis of citizenship in Hamilton. I assess the tools available to citizens to advocate for care, and I show how bioterrorism takes root in people's worldview through a sense of communal vulnerability to disease. I conclude by considering the utility of biosocial groupings as political tools, noting how they cohere and dissolve in changing contexts and how citizens transform uncertainty into daily living.

FLUID FENCES

The Rocky Mountain Laboratories facility is known by several names in Hamilton, Montana: "the spotted fever lab," "the tick lab," or simply "the labs." The word *laboratory* may conjure images of rooms lined with flasks, formaldehyde-filled jars, and Bunsen burners, but the exterior of the building belies this stereotype. Situated in a neighborhood of Victorian houses set on wide, cottonwood-shaded streets, the red-brick facade of RML suggests a turn-of-the-century boarding school. The events that take place within these walls, however, turn it into the known cultural form of a laboratory.[5] Today, the laboratory is secured, in part, by the traditional security technology of a fence—the same type of black steel fence with sculpted but menacing points that frames and secures the White House. Before the fence was built in 1995, RML had an open campus, where high school students congregated in the library to study after school and local farmers strolled into the labs with their diseased chickens in hopes of getting a diagnosis. Now, visitors to RML are stopped at a visitor center, set outside the fence at some distance from the main facilities. Though nominally a public space, the visitor center is a guard station and gateway to the laboratory. Visitors who wish to read the brief history of the lab presented in the visitor center must first show a driver's license and undergo security screening, sacrificing anonymity to even approach the laboratory space.

RML's fence may fend off terrorists, but it also excludes the farmer with his chickens, the curious high school students, and its own employees. Dr. Willy Burgdorfer, famous for discovering the bacterium that causes Lyme disease, was granted emeritus status when he retired in the mid-1980s. Under the new security protocols, Burgdorfer subsequently required an escort to enter the laboratory. Unwilling to work in a place where he couldn't enter "without being accompanied by someone who controls you," he refused to set foot on campus. Acknowledging that there was work that should "stay hush," this soft-spoken scientist worried that the new security measures sent a particular message about the science itself: "If you put police at the entry to the lab, you are saying that you need police protection to make sure [outsiders] don't do things with the equipment. That's no good. That's automatically an invitation to do so."[6]

While the movement of people in and out of the lab is carefully regulated, other things come and go freely. Laboratory work produces byproducts like noise and fumes that are easily dispersed into the community. RML's standing in Hamilton has been tainted by its handling of waste, including toxic specimens and animal carcasses. One lawsuit required the National Institutes of Health, the agency that oversees RML, to pay for the cleanup of a local landfill.[7] Another retired scientist, Dr. John Swanson, now identifies himself as a "neighbor" of the lab. As a neighbor, Swanson has specific complaints about the "shrieking" of the laboratory's incinerator and the air-handling systems installed on top of the buildings, cleaning the air at the cost of noise pollution.

Along with the steel fence and visitor center, this output situates the laboratory as part of a community. In *Laboratory Life,* Bruno Latour and Steve Woolgar sketched the laboratory as a square box into which animals, chemicals and energy are input and from which "articles" are output. Latour used this diagram to map the complex relations within a laboratory that give rise to scientific papers, which are the material

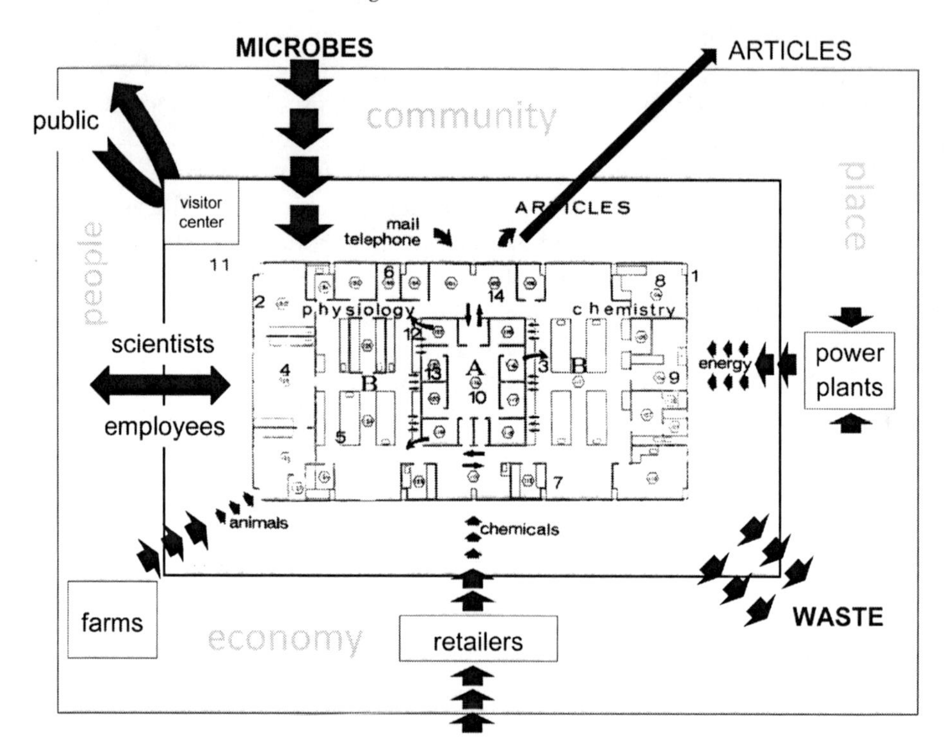

FIGURE 11. Extending outward the flows in a laboratory diagrammed by Bruno Latour and Steve Woolgar in *Laboratory Life* (1979) illustrates how communities produce science knowledge and how laboratories affect communities. Bruno Latour and Steve Woolgar, *Laboratory Life: The Construction of Scientific Facts* (Princeton, NJ: Princeton University Press, 1986), 46.

form of knowledge production. Situating this box within a community, however, would also show outputs of noise and waste, connections between local farmers and the laboratory test animals they supply, scientists moving between home and work, and the draw of energy from local power sources. Supplies, including microbes, also pass through the community en route to the biological laboratory from afar. Documenting the flow of goods into and out of the laboratory elucidates that the steel fence is a mainly symbolic boundary between the laboratory and the community. The inputs that fuel the production of knowledge (energy, supplies, labor) are largely supplied from the local environment, showing the investment of the community in the work of making scientific knowledge. The circulation of agents between the laboratory and community is vital to supporting the laboratory's work, but it also creates risk. The goal of biosecurity is to construct an environment where the necessary materials for science practice can move easily while dangerous items are contained.[8] In a BSL4 laboratory, every flow is regulated, including the flow of scientific knowledge to those who would use the information to inflict terror on populations.

The expansion of the bioscience complex in the twenty-first century has brought more people into physical proximity with high-security laboratories than ever before. The materiality of knowledge production poses a risk not only to scientists but also to those who live near the laboratory space. Citizens demand interventions to mitigate risk, particularly through the control of laboratory spaces. As the first such lab to be built after the terror attacks of 2001, the laboratory at Hamilton was the front-runner in negotiating these risks. Investigating the hundred-year history of Rocky Mountain Laboratories, however, shows that the negotiations over the BSL4 laboratory involve ideologies and community relations that have been contested since the lab's creation. In a sense, the community of Hamilton was formed around a shared experience of disease. Amid changing political ideas, shifting economic resources, and expanding knowledge, a local public-health program of disease management developed into a national science laboratory. The history of RML, including its expansion in the twenty-first century, shows how laboratories exist simultaneously as scientific and political institutions, borrowing the tools and practices of both science and government to claim authority over citizens' lives.

A RIVER OF MYSTERY

> Down the length of a seventy-five-mile valley in Montana runs a river of mystery. . . . A hundred years ago Indians believed that the western side of this river and especially the mountain canyons were inhabited by evil spirits. When white settlers came they observed that, although dwellers on the east side were immune, those who built their houses on the west ran the danger of falling victims to a strange illness.
>
> It began with chills, an aching head, and painful joints and muscles; then a raging fever developed, and a red rash flamed out on chest, back, arms and legs. Rocky Mountain spotted fever was the name that was given to this terrifying and usually fatal sickness.
>
> —*Elise McCormick*, "Death in a Hard Shell," *Saturday Evening Post*, November 15, 1941

The river that bisects the Bitterroot Valley once marked an inexplicable barrier between life and death. At the end of the nineteenth century, a deadly disease emerged in the Bitterroot. The mysterious plague, known as black measles or mountain fever, killed four out of five people who contracted it. The disease cast a pall over the valley and was identified as "one of the biggest factors in retarding the growth of this part of the country."[9] Although the river was shallow and regularly traversed by humans and animals, the infection prevailed only on its western shores. As people watched their neighbors succumb, their community was further divided by fear and suspicion. As one resident said of property on the west side of the river, "I wouldn't have it and live on it if they would give it to me—it's in the Spotted Fever District."[10] The chilled streams flowing from the western mountains seemed the only plausible carrier of the fever until a pathologist in Missoula proved that a pin-sized wood tick, rising from the earth in the spring in search of a blood feast, was the carrier of the deadly microscopic bacterium *Rickettsia rickettsii.*

Though early Montana settlers did not know the source of the spotted fever, they created a local knowledge of the disease delineated by geography. Just as modern citizens recognize malaria or yellow fever as diseases of the tropics (and take precautions when entering those areas), Bitterroot residents ordered their world and modified their behavior according to their understanding of the land as half diseased, half healthy, with the river running down the middle. Although later work of scientists in the Bitterroot Valley produced new ways of knowing the disease, its natures, and its movements, conventional geographic understandings persisted, influencing both control measures and public sentiment for decades.

While the river created a territorial division with respect to the disease, it also bound the population through their shared susceptibility. The belief that the river isolated the fever's source from the town created a safe space for citizens, who claimed territory on their side of the river to "assure a stable sense of community" and bring what Charles Maier calls "identity space" into alignment with the "turf that seems to assure physical, economic, and cultural security."[11] Here, security came from a geographically specific nature that was free of disease. Thus, the very notion of community is partially predicated on the belief that proximity to other community members in a chosen landscape increases social and economic benefits without compromising physical security. Governance becomes acceptable to keep this relationship in balance.[12]

From 1903, the State of Montana sent a new specialist each season for a decade to study the problem. It transformed the Bitterroot Valley into a laboratory with the objective of understanding spotted fever. Eventually, the entomologist Robert A. Cooley set up a semipermanent laboratory in an abandoned log cabin on Sweeney Creek, where he and his three assistants studied the life cycle and habit of the Rocky Mountain wood tick, undressing every two hours for a head-to-toe tick inspection.[13] Their research produced knowledge of an insect whose eight legs do very little to move it about in the world. A tick may pass an entire season on a single blade of grass, waiting to attach itself to a passing mammal and feed on its blood in order to produce offspring and pass on its genetic code. Thus, the movement of the tick on the landscape follows the movement of its mammalian carriers. Wherever an animal can walk, a tick can travel, which means a tick can cross the Bitterroot River or hitch a ride on a scientist traveling home from work. The perceived isolation of spotted fever on the west side of the river was more a reflection of human behavior than of tick or microbe movement: where people lived they cleared vegetation, but on the other side of the river they walked and rode through trees and brush, thus coming into contact with ticks and *Rickettsia*.

In July 1912, Thomas B. McClintic, a scientist detailed from the national Public Health Service, fell ill as he was packing for his return to Washington, DC. His health declined on the cross-country train ride, and he died just hours after arriving home. Major newspapers headlined his death, and national interest grew, accompanied by increased funding for public health interventions.[14] The early records of Montana's new Entomology Board show the employment of a range of traditional public health strategies, such as quarantine and species extermination, followed by innovative land-management techniques. Only after ten years of tick control proved unsuccessful in eradicating the fever did the board call for more scientific study of the disease and its transmission.[15]

Though public health officials promised to "completely . . . clean up the entire Bitter Root valley," they often met with opposition from the community, because their programs asserted state control over land and livestock.[16] The first arm of the state tick-control program was the "destruction of adult ticks on domestic animals," which was accomplished by driving cattle through an arsenic-laced dip.[17] The law required ranchers to bring their own animals to the vat. Those who refused had to show they had hand-picked ticks from their animals, otherwise their farms were subject to quarantine, an authority granted to the Board of Entomology by a 1913 law.[18] Both dipping and hand-picking required substantial time and labor, perhaps accounting for the program's unpopularity among locals. When John Dunbar was arrested on June

FIGURE 12. Public health workers in the Bitterroot Valley run cattle through an arsenic solution to kill ticks. Protesters destroyed a dipping vat near Stevensville in 1913. Photo: Rocky Mountain Laboratories, National Institutes of Health.

16, 1913, for destroying a cattle-dipping vat near Stevensville, the community celebrated with chants of "Dunbar! Dunbar! He done it with a crowbar!"[19] The deaths of livestock and eventually of two vat workers brought an end to the mandatory arsenic baths.[20] More than a demonstration of the force of law and the power of the rising public health agency within the community, the dipping program served as a visible show of state involvement in the health of citizens and animals.

As dipping livestock became increasingly unpopular, the state concentrated on managing rodents, which can become carriers of the disease when ticks in the nymph phase first feed on their blood. For a fee of five to ten cents per acre, state workers spread poison on farmers' land, enticing hungry ground squirrels when they emerged from hibernation.[21] This method, however, failed to reach the vast expanses of uncultivated land that also harbored the wood tick. In the face of repeated failures, the board concluded that the management of livestock movement in the landscape was the key to avoiding infection. If stock could be kept away from tick-infested areas in the spring, the spread and survival of the tick might be inhibited. The board was, however, apparently unwilling to fight the legal battles that would force cooperation between landowners, the valley's numerous part-year residents, and government agencies like the forest service, favoring instead visible acts like cattle dipping and spreading poison. Moreover, citizens did not cede full authority over the health crisis to the state. Simple acts like staying on one side of the river expressed individual autonomy. As one resident explained, "If I believe that it's the snow water, I can carry my own well water with me and not worry. If I had to believe it was ticks, I'd just be scared all the time."[22]

After five years' work, the State Board predicted an optimistic and healthy future for the Bitterroot: "Based on good, substantial reasons, the residents, both on the farms and in the towns, have a greater confidence in the future of the valley[, and] there is less apprehension concerning the disease."[23] Through its work on tick control, the government had also increased its authority over the land and produced an understanding of insects as vectors of disease, propagating the perception that citizens' health was connected to their natural environment. The most enduring outcome, however, was the presence of the scientists. The work on spotted fever had brought leading scientific researchers into a place where most people drew on the resources of the land for their livelihood. In an unanticipated twist, the devastating bacterium made national science research a major factor in the economy of the valley.

When Ralph Parker came to assist in the spotted fever effort, he had shunned Cooley's rustic lab and set up in a woodshed behind his house. Parker's early work centered on quantifying the occurrence of the disease through a tick census, sweeping large flannel flags along animal trails and then counting the ticks clinging to the fabric. Back at the woodshed, he sorted, tested, and catalogued the specimens, using live guinea pigs to identify ticks carrying *Rickettsia*. In this makeshift laboratory, Parker transformed ticks from matter into data.[24]

While testing the infection rates from ticks in various phases of development, Parker stumbled on evidence of immunity. Guinea pigs did not contract spotted fever when they were bitten by young ticks that carried the bacteria but had not taken the blood meal that would nourish them to adulthood; later, these animals demonstrated immunity when infected with the disease. After making this discovery, Parker focused on the development of a spotted fever vaccine, turning a field-based study of the landscape into a laboratory experiment involving test subjects and equipment.[25] The promise of a vaccine and a particularly deadly spotted fever season in 1921 prompted the U.S. Public Health Service to enlist Parker as a federal employee, allocating twenty-seven thousand dollars and a team of scientists from Washington, DC, for vaccine research. (Field control work still fell under Montana state jurisdiction.) Additional funds were raised by the local chambers of commerce. Parker's first task under the new administration was to find a suitable laboratory.[26] He settled on an abandoned school building northwest of Hamilton, which he crammed with collections of ticks and rodents and vials of serum. Because vaccine preparation required the handling of infected ticks, grinding them down with a mortar and pestle into a virulent serum, laboratory workers came into daily contact with deadly spotted fever. A 100 percent fatality rate among lab workers who contracted the disease produced a public image of the "schoolhouse lab" as a dangerous and unsafe site, full of cracks and crevices where infected ticks could hide.

Despite the limitations of the facility, Parker produced a spotted fever vaccine in 1925. Over the next three years, more than two thousand individuals in the Bitterroot Valley received the vaccine, a demand which challenged the lab workers grinding up ticks for vaccine in already cramped conditions.[27] In 1927, the Montana State legislature appropriated sixty thousand dollars "to build and equip a laboratory for the purpose of conducting experiments and other work for the control of Rocky Mountain spotted fever and insect-borne diseases."[28] The Board of Entomology selected twenty-eight acres of land adjacent to the city of Hamilton as the site for a new laboratory. Significantly, the proposed laboratory would sit on the eastern bank of the Bitterroot River.[29] The previous three laboratories were located west of the river, on land already presumed to be tick-infested. The new lab would bring ticks over the river, violating the geographic barrier that culturally defined the risk of the disease. Though scientists had discredited the myth of the river barrier, the sway of public perception was displayed in June 1927, when Hamilton residents filed suit in the Fourth District Court to contest the selection of the site. The cultural knowledge of disease risk in the Bitterroot Valley, which had taken root long before scientists remapped the territory of spotted fever, held firm in the citizens' worldview. The proposed site located the work of disease scientists in the center of the community, and the laboratory now bore the responsibility of proving that the circulations of people and matter would not put citizens at risk.

In the 1927 lawsuit against the proposed site of the new lab, the plaintiffs argued that they were not "complaining of the law for the building of this laboratory . . . [for] there has been wonderful work done there for this community."[30] Nearly eight decades later, when the Coalition for a Safe Lab, Friends of the Bitterroot, and Women's Voices for the Earth filed a suit in U.S. District Court against the National Institutes of Health, they were accused of being "against" RML. Called an "unpatriotic, leftist fearmonger," Mary Wulff grew tired of explaining that the lawsuit did not seek to stop the building entirely.[31] Jim Miller, with Friends of the Bitterroot, claimed, "it was never our intention to stop the lab. We could have delayed it, but stopping it was never our intention. It was never a possibility that the lab would not be built."[32] However, both legal cases were formulated on the idea that Block 19 in Hamilton, Montana, might not be the best place for the type of science being done in the labs. In both cases, scientists and their lawyers described the laboratory as a safe and carefully managed space, and residents demanded proof that the laboratory work did not threaten their health. Examining the arguments presented in two courtrooms nearly a century apart gives insight into the deeply instilled values that underpin the formation of a community around citizens' claims to health and biological security.

FEAR AND THE DUE PROCESS OF LAW

The decision to take the government health institutions to court reflects a multigenerational presumption that citizens' biosecurity is the purview of government. One of many venues for the negotiation of biological rights, judicial systems provide means for citizens to seek redress for biological injury. Adriana Petryna's investigation of how Chernobyl victims petitioned for social welfare through the legal systems shows how legal action established categories of citizenship and mechanisms for creating a "stark order of social and economic exclusion."[33] Legal negotiations of biological citizenship create collectives through precedents. In Hamilton, however, citizens did not seek redress for injury but sought to deter future harm. The litigation demanded specific government action that would enable citizens to endure the biological risks of everyday living. Still, as Petryna observes, "daily life is characterized by overwhelming uncertainty and unknowability. It is in this social, scientific, and legal arena that defining and acquiring a biological citizenship takes on central interest."[34] Hamilton residents took their concerns for the security of daily living to courts of law, asking legal authorities to calculate the risk of the laboratory and define the reach of governments to manage it.[35]

"The building of a laboratory is a violation of the rights of citizens without due process of law," stated the plaintiffs in their opening statement on July 27, 1927, arguing that an act by the legislature is illegal if it interferes with individual rights to safety, health, and happiness in a clean and healthful environment.[36] In its deliberations the court would weigh the rights claimed by citizens who lived near the laboratory against the rights of a broader group of citizens to potentially gain better health through the study of disease. With respect to the laboratory, the court had authority to define the place as either a threat to citizens or a contained space that did not pose an external threat great enough to violate human rights. The case hinged on the question of whether the law should regulate how laboratory work was done—and where—for, as the defendants claimed, "it is hardly within the province of a court of equity to enjoin the operation of a laboratory the work of which constitutes an obviously innocent and harmless purpose."[37] The plaintiffs had to prove that it was the responsibility of governments to moderate the risks of laboratory work and demonstrate the potential benefit of the research to society.

Testimony ranged from heartfelt pleas on behalf of innocent children and chilling accounts of deadly laboratory accidents to commendations of scientists who saved lives and descriptions of "tick-proof" facilities. Both sides struggled with the task of quantifying fear and bringing it into the realm of intervention. The case presented freedom from fear as a human right, a state of being to be protected by the court. For example, when Dr. G. A. Gordon took the witness stand in 1927, he was identified as a physician who had practiced in Hamilton for eighteen years and who lived a block and a half from the proposed laboratory site. Though Gordon mentioned his experience in treating spotted fever cases, the majority of his testimony was directed at establishing that the community feared the new laboratory. He testified:

> Well, the general impression of the people of that vicinity is one of fear; fear of the presence of the laboratory; fear of its being built there. We have a laboratory on the west side and they know how it is conducted, and the general impression on even that community, where it is, and even myself is one of disgust. And the people of this community, where they are proposing to erect this building, they can't understand why anybody or any body of men would want us, would ask us, to have it near us. Even though there were no danger from contamination; even though the grounds and building were conducted as it should be. Yet in the minds of everybody in that neighborhood there is fear to such an extent that they do not consider they can live in ease and peace of mind with a laboratory of that kind in the vicinity.[38]

In redirect examination, Gordon listed by name individuals he had spoken with who had expressed fear, after the cross-examining attorney presented his list of those who were unafraid. These exchanges aimed to draw boundaries around fear, one side contending that fear came from living near the laboratory, and the other arguing that fear was needlessly cultivated by association with fearful people and had no basis in lived experience. This line of questioning also put on the court record names of individuals who experienced this fear and who used it to claim they had been wronged.

The defense worked to establish that people's fears were unjustified or unavoidable. In one cross-examination, the defense lawyer referred to the folk wisdom that the disease was carried in snowmelt, asking the witness if he thought those rumors were groundless:

A. I would say so.

Q. You don't think that drinking the water causes Rocky Mountain spotted fever?

A. I would hardly think so.

Q. You have heard people say that is the case though?

A. I have heard it, yes.

Q. That did not make you afraid, did it?

A. Well, not exactly.

Q. Mr. Hagens, don't you exercise your own judgment when you hear rumors about the water of the Bitter Root river?

A. I certainly would.

Q. Would not you exercise the same judgment with reference to these rumors with reference to possible danger from the tick laboratory?

A. The dangerous element there is so much more dangerous than the drinking of the Bitter Root River water. . . .

Q. You would be afraid of rumors no matter whether they were groundless or not—rumors from the tick laboratory?

A. About the ticks, I certainly am afraid.

Q. I am trying to find out now, whether this is your judgment or just being afraid.

A. I would worry a great deal over such rumors as to whether they were true or not. . . .

Q. If you found out that this was a cement building, tick proof, rodent proof and vermin proof, that would allay your fear to a considerable extent, would it not?

A. Well, somewhat. Not very much.

Q. You would still worry about the rumors, even though you found they were groundless?

A. Anything as dangerous as that, I surely would.[39]

In claiming the right to rely on his own assessment of danger, the witness establishes a hierarchy in which this assessment outweighs factual information, particularly when risks are high. Such claims establish fear as an inexplicable entity, naturalized in the environment. If knowledge is the antidote to fear, but no knowledge is strong enough to overcome the fear, no action can be taken to eliminate fear. The defense rests—and throws up its hands: "It is not humanly possible to answer the person who says he is afraid and cannot give a reason for his fear."[40]

Next, the plaintiffs paraded a handful of janitors and caretakers through the witness stand to testify to the numerous accidents and incidents at the schoolhouse lab, challenging the alleged security of the laboratory space and posing the materiality of the building as a biological threat. An animal handler told of a goat that escaped its pen and a porcupine that climbed through the roof. Other workers told of broken bottles and tick sacks with gaping holes in the bottom, and of scientists scrambling about the yard with white cloths, attempting to flag the ticks that had escaped from the laboratory. A janitor recounted how his son had contracted spotted fever while visiting him at work, a custodial job that included inoculating guinea pigs when the lab was short of staff. These testimonies painted a picture of scientists running a haphazard operation, unable to control the larger animals, let alone the ticks, and exploiting untrained laborers to do deadly work. Janitors and caretakers were portrayed as mopping up messes left by careless scientists, as common citizens fulfilling the role of "sole protector of the community."[41]

The defendants contended that it was not the scientists who were to blame for these incidents but the poor infrastructure of the laboratory. They argued that a new, tick-proof, fireproof, rodent-proof laboratory would eliminate these risks, a promise met with skepticism by the residents:

> *Q.* If this laboratory is properly constructed, and according to the best scientific knowledge of the day, and if it is operated according to the best knowledge for the protection of laboratory workers, you do not think there is a real danger of infection being carried.
>
> *A.* There is a danger. Why would it be any different where they are working now, and where they will work[?]
>
> *Q.* You think that the new laboratory cannot be so constructed as to minimize the danger that may exist from the operation of the present laboratory.
>
> *A.* Well, it may minimize the danger, but it will not exclude any possibility. You will have a case of spotted fever out there, then you will wake up to what we are saying.[42]

Like the scientists at Fort Detrick, the defendants argued that human behavior creates vulnerability, but that the laboratory could be built so as to prevent harm from human error. When the issue arose again in the following century, the director of RML recognized that he had to tackle the question head-on. In describing the layers of security of the BSL4 laboratory, Dr. Marshall Bloom explained that administrative oversight, training, and standard operating procedures (SOPs) become more stringent as the need for safety increases, and employees who ignore safety protocols risk losing their clearance to work. The laboratory regulates its human scientists by categorizing behavior, writing detailed prescriptions for safe behavior, and punishing behavior considered unsafe.[43] In contrast to the terse warning scrawled on a sign above the old schoolhouse lab, "Enter here at your own risk," RML's multivolume binder of SOPs shows close attention to managing the human causes of risk. The terrorist attacks of the twenty-first century further complicated these concerns. Even if the laboratory could guarantee that it had designed a space that would eliminate all risk due to human error, could it guarantee that the laboratory was secured against external breaches such as a plane crashing into the building or a scientist deliberately removing specimens?

In the 1927 case, after three days of testimony, Gordon, Hagens, and the other plaintiffs failed to convince the judge that proximity to the laboratory violated their human rights by putting them at risk of infection. The judge dismissed the case on the grounds that a new facility would contain insects more effectively and cited the precedent of laboratories in urban centers that "are not regarded as dangerous to health nor as a nuisance."[44] Throughout the hearing, the lawyer for the defense sought to discredit the plaintiffs' witnesses, who were primarily citizens of Hamilton residing near the lab site, questioning the authority of a doctor to address scientific matters outside his medical practice or that of a resident to claim any knowledge of property values. The citizenry were presumed ignorant, and their means of producing knowledge dismissed as rudimentary compared to the scientists'.

Three outcomes of this court case bore on the future case at RML as well as on broader discussions of how citizens navigate the legal system to advocate for biological rights. First, the defense convinced the judge not to intervene in a case where the plaintiffs were "seeking to enjoin through the imaginary future condition." Those advocating for protection must convince the courts that the risk of future biological harm warrants directives in the present. Second, the decision clarified the court's jurisdiction over the laboratory space, affirming the authority of the judicial branch in regulating the laboratory on behalf of the community. Third, in 1927 the judge used a relative assessment of danger on the barometer of tick-related harm, comparing the risk posed by the laboratory to the existing environmental risk of living in Hamilton. The right to health did not include the right to not be exposed to diseases one could "naturally" pick up walking down the street, a danger that "cannot be eradicated by any known means." By living near an area where Rocky Mountain spotted fever occurred naturally, citizens had already relinquished their rights to live without the disease. Moreover, the work of the laboratory to eradicate the disease was primarily intended to benefit the local community, an "obviously innocent and harmless purpose." The defense lawyer proclaimed that the work on ticks at the biolab was far safer than the work of scientists who "handle organisms too small to be seen by the human eye. . . . [T]here is far greater danger from the handling of these than from the handling of an insect that can be seen by the human eye."[45] Ironically, by World War II, the laboratory was conducting research on microorganisms like typhus and yellow fever. When RML announced plans to bring BSL4 pathogens to Hamilton, scientists were already studying dozens of diseases that could not be found in the Bitterroot Valley. By definition, BSL4 pathogens do not occur locally, meaning that the 2004 court case was predicated on a different politics of nature, in a situation where citizens were proximate to germs in the laboratory that they would not encounter in their local environment.

THE HEART OF COMMUNITY

International interest in insect-borne diseases and a steady supply of national funding brought more scientists to RML: in 1941 it employed ninety-seven people.[46] Over time the laboratory became more closely integrated into the Hamilton community. Partly because of its isolated location, RML obtained most of its supplies locally. Maurine Hughes remembers her brother raising guinea pigs in cages behind their shed, which he sold to the laboratory to support the family. When an egg-based vaccine replaced the tissue-based spotted fever vaccine, there arose an egg economy, with residents selling home-grown eggs to the laboratory for vaccine production. Though RML was not a diagnostic lab, scientists developed connections in the community. When a farmer thought her chickens might have tuberculosis or a forest ranger found a bighorn sheep that had met with a peculiar death, scientists might come into work to find a sheep skull on a work table or a dead chicken strung over the back of a chair.[47]

In an interview, Burgdorfer contended that these connections between laboratories and their communities are critical for "good science": "When you hear of an infection, you can't just drive your car into an area and solve the problem. You have to establish a contact. You'll learn more about the area by contact with a farmer than with a scientist." The Swiss-born scientist, once president of the local bowling association, worried that today's scientists were coming to Hamilton to work in the labs but not to be a part of the community. As we conversed in a booth at the Coffee Cup diner, people waved hello or paused to chat as they passed by, inquiring after his health, family, and life during retirement. Burgdorfer laughed at the memory of friends coming to him for advice regarding every headache, sneeze, and fever: "I would tell them, 'I am not a physician, so I will give you a layman's diagnosis: it is the end of tick season, we had a lot of Colorado tick fever this year; it is not Rocky Mountain spotted fever, so it must be viral. You will recover by yourself.' "[48]

Burgdorfer and Dr. Bill Hadlow met for coffee every Thursday for almost fifty years. Both were renowned scientists, Burgdorfer for his discovery of the pathogen that causes Lyme disease, *Borrelia burgdorferi*, and Hadlow for groundbreaking research on prions and bovine spongiform encephalopathy, or mad cow disease. These men described RML as a collegial workplace, recalling their colleagues and technicians by name and fondly reliving laboratory lore. They talked about their era of science as a time of asking questions, following threads, and formulating new problems—an approach they feared might be disappearing in an era of electron microscopes and scientists who are trained to do, not ask. They narrated their careers with tales of people and the peculiarity of doing microbiology in rural Montana and illustrated how the laboratory became the cultural heart of the Bitterroot Valley, enduring long after the immediate threat of spotted fever abated.

"After the 1920s, there was no reason for the labs to be here," said Swanson. By the end of that decade, the spotted fever disease no longer posed a lethal threat to the community, but a small economy had been built around disease control. People earned a living running dipping vats, flagging areas for ticks, or caring for animals at the laboratory. The construction of the new laboratory drew close to a million dollars into the community in the 1930s.[49] Every scientist who came to Hamilton needed a support team of technicians and caretakers, most of whom were hired locally. Where decades ago one might have said that Hamilton needed scientists to save lives, city planners now say that the community needs RML to save its quality of life. A 2002 economic needs assessment of the Bitterroot Valley urged the county to focus on assets it already had, such as RML, while maintaining the "high quality natural environment and setting in the Bitterroot Valley [as] one of the area's key economic advantages."[50]

In June 2008, the Ravalli County Economic Development Authority announced plans to build a $1.6 million small-business incubator in Hamilton, designed to "grow" and "nurture" community-oriented businesses in the Bitterroot Valley.[51] The Ravalli Entrepreneurship Center (REC) opened in 2010 with workspaces encircling a courtyard to encourage meeting and intermingling, creating a place for networking as well as spaces to run experiments. Reminiscent of RML in the days of Burgdorfer and Hadlow, this space emphasizes the advantages of proximity to other scientists for sharing ideas and resources. Community planners hope that expanding the bioscience economy will bring prosperity without cutting down the forests or polluting the clear skies that draw people to the Bitterroot to begin with. Julie Foster, head of the REC project, is quick to point out that this type of development is possible only because of the presence of Rocky Mountain Laboratories, which provides a base of human capital.[52] Most rural communities would not be able to attract scientists. Hamilton, however, already boasts one successful offshoot of the RML powerhouse: a glass-walled prefab factory on the north edge of town, the site of GlaxoSmithKline Biologicals (GSK), established in 1981 by four former RML scientists. GSK claims to have brought three hundred jobs and added $3 million to the county's tax base.[53]

The new biology generates new economic activity, binding contemporary capitalism to biotechnology and changing the materiality of life through commodification and economic production.[54] The life science laboratory sits at the center of a dynamic economy, its circulations part of broader capitalist structures and its materiality part of a contiguous geographic community—and vice versa. In conceptualizing the modern laboratory and its value and risks, one cannot ignore such circulations, which are the basis for a community's shared identity. When citizens advocate for the elimination of collective biological risk, they are acting on behalf of a community that is defined, in part, by the shared experience of the local market and geography.

BIOSOCIALITY AND THE GOVERNANCE OF RISK

> One side would say that that the probability of something happening is 0.0000037. From a scientific view, that's bombproof. But on the other side, the emotional side, that number is a measurable thing, something that could happen. The argument raged because neither side could hear each other talking.
>
> —Russ Lawrence, Hamilton resident, July 23, 2008

Paul Rabinow uses the term *biosociality* to refer to the situation in which new biological materialities bring about new social orderings.[55] He argues that as people discover new "truths" about their bodies, they form groups and identities around these biological understandings of themselves. These biosocial engagements produce subjects whose corporeal experiences substantially inform their ways of living in the world. In the modern biosocial world, alliances formed around biological identities engage in collective practices that remake the world along the lines of biology. While scientists are producing new truths about human biology, they are also creating a world of complex nonhuman natures and materialities that further shape the biological experience of community.

Considering the microbe in opposition to human political aims must take account of the ways the microbe contributes to the constitution of the human body, a microbiome moving in a microbiosphere. On the scale of the microbe, biosociality explains how the work to police immunity, contagion, and vulnerability to germs changes how people live in the world. More than just shared experiences, the regrouping of society through biosociality raises questions for democracy and creates new spaces and mechanisms with which to contest the biological politic.[56] These biological alliances bring health into the political and economic domain, placing individual biology at the center of citizenship.[57] As people engage with each other on the basis of biology, as Rabinow observes, life "is embedded throughout the social fabric at the micro-level by a variety of biopolitical practices and discourses."[58]

Theorizations of these biosocial groupings have been based largely on of biological states of difference, such as genetic or other medical conditions or environmental harm. Such groupings have focused on issues such as access to care, the formation of support groups, or attaining redress for injury.[59] In biological collectives created through bioterrorism, the subjects in alliance have not suffered shared injury. Their collectivity is not based on common understandings of their bodies as they are now but on shared perceptions of how they might be in the future. Biosociality in the age of bioterrorism involves the production and consumption of information to bring about shared ways of knowing biological threats and collective ways of advocating for a common biological future.

Public discourse about the Integrated Research Facility at RML hummed with quantifications, articulations, and speculations, of biological risk. As the local bookstore owner Russ Lawrence remembers, people found diametrically opposite meanings in the same calculations. While some focused on probability, others pondered possibility; some shared numbers, others shared stories. Collectively, the community tried to assess the potential for pain and suffering from an imagined future event. They used a variety of political instruments, from environmental impact statements and town hall meetings to legal challenges and community organizing. A close reading of the scientific, legal, and social criteria for citizenship formulated by this activism exposes complex struggles by citizens negotiating a world of unspecified biological risks. Although the laboratory remained at the center of this negotiation, it was a shadow puppet representing the many risks enlivened by modern natures.

In their refusal to accept any measured assessment of the biological threat, opponents of the lab expansion rendered risk incalculable. In such circumstances, Sheila Jasanoff observes, risk can no longer be managed but must be governed. A governing approach to risk requires attention to the social institutions and myriad interspecies encounters that shape people's perceptions and experiences of risk in order to find an intervention. Because governance garners power through the aggregation of preference, experience, and beliefs beyond the calculation of science, it allows imagination and conjecture to enter discussions of risk. By acknowledging the limits of expertise, risk governance might be better able to incorporate the past experiences and perspectives of the public than a management strategy entrenched in probabilities. To be effective, this strategy for mitigating risk demands continuous collaboration with the authorizing public while maintaining a foundation of trust and transparency. Jasanoff contends that if done well, policing risk through systems of governance builds a resilient population as well as "political environments in which risk morphs into reality," offering more hopeful futures to engaged citizens.[60]

The events surrounding the RML expansion, however, present a less rosy picture of governance, in which people's efforts to participate in cultural articulations of risk were initially ignored and then repeatedly undermined by a technocratic system determined to calculate risk through scientific method. Citizens' desire for transparency could not be satisfied by the governing entities; miscommunications led to confusion and anger in a conflict that raged for five years. These years of negotiations elucidated the need to rethink how biological risks can be governed in a time when genomic science continually produces new ways of knowing life itself. In Hamilton, the alliance of citizens were unable to counter the powerful calculations produced by the science establishment that demonstrated low risks. The claims of individual and collective future harm faded beneath the high value ascribed to the science industry locally and within the massive national security complex.[61]

Between 2002 and 2008, citizens made a case for their vulnerability to biological threats and called on political systems to govern biological risk. People use public institutions to negotiate modern natures. These institutions in turn delineate the value of human lives. By harnessing the social institutions that regulate natures in order to manage the risks presented by new biological forms, this work produces the biosecurity state. Further, the citizens in Hamilton demanded that governments evaluate environmental risk in terms of new microbial dangers, demonstrating that people know nature differently in the post-9/11 world. The performances of biological citizenship staged in Hamilton were both individual and collective. The crisis in Montana shows that even when united in a biosocial group, people struggle to articulate their personal and collective experiences of biological risk persuasively in a public discourse dominated by the growing national security complex.

THE ENVIRONMENTAL REVIEW OF RISK

Now little bitty Hamilton—little picturesque Hamilton, the Bitterroot Valley—is facing a nightmare they have never seen before. How do you contain a nightmare? It is impossible. Foolish men have throughout time stated that it is possible to contain, it is possible to keep ships from sinking, and every time they are proven wrong. Ships do sink. And biosafety hazard level four containment areas do get breached. It happens. And that's just, that's just too much for this little community. It's too much.

—Matthew Lemax

If there are risks I want to hear about them—in detail. If there are alternatives, I want to hear about them—in detail. I want to be able to make an intelligent decision in my own heart about this lab, because I know you all do wonderful work over there. But I don't believe that we've been treated as the intelligent people that we are in this community. And I don't like being condescended to. I don't like snide comments back on my comments. I made a comment in another meeting, "Is this a done deal? Should we just all go home? Is it already a done deal?" And I was told I hadn't been listening. I have been listening. I have been paying attention. But we're not getting answers. And I will go back to my very first comment, the very first meeting that I went to. I can tell you one thing: People back in Washington, DC, in Bethesda, Maryland, do not give a damn about people in Hamilton, Montana. There are only a few of us, but they don't care about us. And I tell you, you're naive if you believe that they care about you. They don't. So, my question, Marshall: is this a done deal?

—Joan Perry
Comments made at the first public hearing for the draft environmental impact statement for the RML Integrated Research Facility, June 2003

How do you measure the environmental impact of a laboratory? Under the National Environmental Policy Act of 1969 (NEPA), any building project involving federal funding is subject to environmental review, a process designed to expose not only the environmental but also the social and economic impacts of infrastructure development in the United States. When the National Institutes of Health initially proposed the BSL4 expansion at RML, they assessed the affected area to be one hundred thousand square feet, roughly the footprint of the proposed building. The project timeline assumed that the agency would do a basic environmental assessment, which would lead to a "finding of no significant impact," allowing construction to proceed without undertaking a longer and more rigorous environmental impact statement (EIS). Recalls one activist, "There was literally no acknowledgment [at the time] from NIH that this lab would have any more potential impact on the community than a large office building would."[62] Yet the proposed BSL4 laboratory space, housing deadly microbes, rendered the space unlike any other facility to be built under NEPA guidelines. It was evaluated against new standards of environmental risk, which increasingly focused on worst-case scenarios and struggled to find mechanisms to assess the reach of microbial threats. By demanding a full EIS, the community demanded that the NIH rationalize all the ways that the building, its systems, and employees interacted with the physical and social environment.

NEPA procedures require opportunities for public comment. In public meetings about the IRF, both parties discovered the difficulty of identifying, labeling, and addressing their respective assessments of risk as "fear," "worst case," or "safety." Citizens expressed frustration with

FIGURE 13. Integrated Research Facility at Rocky Mountain Laboratories. Photo: National Institutes of Allergy and Infectious Disease, National Institutes of Health.

the NIH on many levels—for treating them as ignorant, unimportant, or inconvenient; for withholding information; for being careless; and for discounting concerns expressed in public and private. Supporters of the lab accused opponents of irrationally trying to impede the process, hurting the economy of the valley, and disrespecting their neighbors who worked in the lab. Through the NEPA process, RML and NIH became more skillful at directing public comment and, in the end, institutionalized the EIS as a mechanism for assessing the impacts of BSL4 laboratories, a model likely to be adopted in the future construction of biolabs.

Alex Gorman, who became involved in the RML debate through her work with Women's Voices for the Earth (WVE), believes that the NIH's timeline collided with Hamilton citizens' needs for information and conversation before breaking ground. The lab expansion was driven by the momentum (and money) surrounding national security after 9/11, and the political climate demanded action from the science community to prepare the country for terrorist threats. The first public presentation about the BSL4 laboratory, in February 2002, conveyed the expansion as an opportunity for Hamilton to do its part for national biosecurity and did not conceal the fact that the expansion was intended as a facility for studying "bioterrorism agents" and "diseases caused by the intentional release of pathogens into human populations."[63] The proposal addressed security concerns in vague and presumptive terms, suggesting that citizens could trust the laboratory to look out for their safety. But people living in a world where the unimaginable played out on every major broadcast network in real time were unable to trust the lab when it said the building would be secure.

At the time of the annoucement, RML was on the verge of opening a BSL3 expansion, which was ratified in the late 1990s with an environmental assessment and no public opposition. From an architectural point of view, the difference between a BSL3 and a BSL4 facility is incremental. Less than a decade later, however, the world had changed, and citizens demanded that biological events be scrutinized to the full extent allowed by public process. "Given the timing, people were pretty worried about things like planes that might crash into the lab, etc. It didn't take much for local residents to think up some pretty scary worst-case scenarios. And the response from the NIH, who apparently had not put much thought into worst-case scenarios, was to ask the public to trust them: they knew what they were doing, and it was all going to be very safe. That didn't sit well with a lot of folks—so the whole project got a lot more attention and controversy than the NIH ever expected."[64] NIH would have to prove the security of the facility to the public through definitive, specific, and active measures.

An early and oft-repeated citizen complaint was that the BSL4 expansion was presented to the public as a done deal, discrediting a fundamental principle of the NEPA process: that sometimes the most favorable outcome is to take no action. Even when National Institute of Allergy and Infectious Diseases director Anthony Fauci announced on September 12, 2002, that the agency would conduct an EIS, his statement assumed that the public comment period would alleviate concerns without altering the outcome: "I'm totally confident that the people of the Bitterroot Valley want to be a part of what is happening here. In the same breath, we want to make sure they are a part of the process and comfortable with the lab."[65] Fauci assumed Hamilton citizens' political interest in the lab while issuing an authoritative assurance that the infrastructure would be secure.

Throughout this process, Hamilton residents found new ways to speak out for their biological rights. Groups like Friends of the Bitterroot and WVE included the lab issue in their environmental advocacy platforms. Mary Wulff's Coalition for a Safe Lab staged meetings parallel to the NIH-sponsored gatherings with the goal of enabling citizen participation in the NEPA process. At these meetings, worst-case scenarios and security questions were discussed instead of being dismissed as fearful rhetoric. These discussions also defined the contexts of the argument, locating biological risk as an environmental concern, a local problem, a women's issue, and a matter of laboratory safety.

Around the same time it began the EIS process, RML established a Community Liaison Group, part of the lab's strategy to "be a good neighbor." The CLG included laboratory officials and representatives of various community sectors, such as the fire department and the Chamber of Commerce, selected to facilitate the flow of information between the lab and community members. When the protest against the lab subsided, the CLG endured. In its interactions with the community through the CLG, RML is the gatekeeper of information. One CLG member pointed out that information seemed to flow downhill from the lab, with a lot of propaganda and without much sincere discussion. "They put on the face that they want the community to see—it's nothing but a PR apparatus. They are going to say what will calm you down, until you become accepting of it. New ideas are tough to swallow."[66] RML had little at stake in forming this community group: through the existence of the CLG, the laboratory could promote itself as community-minded while instituting a system that accepted input only from a faux representative council. The power relations were skewed, for CLG members ultimately had no authority to enact change within RML or NIH. By establishing the CLG, however, RML acknowledged that the laboratory exists in a community with which it has a responsibility to communicate in some structured way. Bloom stated, "What we do here is a privilege, not a right—and one given to us by the taxpayers."[67] The CLG also formalized an understanding that the laboratory had to account for its role in protecting the safety and well-being of its neighbors.

The establishment of the CLG brought the community more formally into the scope of the EIS.[68] For the next two years, the NEPA process framed discussions of the IRF. The public comment and EIS documents inscribed the varied sentiments floating around the lab and in the community into a ritualized exchange. The NIH would present a draft EIS; the public would respond; NIH would revise, and the public reacted again. The documents and hearings proceeded as follows:

EA scoping meeting	July 2002
Draft EIS (DEIS)	May 2003
Public comment period, meeting on DEIS	June–July 2003
Supplemental draft EIS (SDEIS)	December 29, 2003
SDEIS comment period and public meeting	January–February 2004

Final EIS (FEIS)	April 2004
FEIS comment period	May–June 2004
FEIS released	July 26, 2004
Lawsuit filed to request new EIS	July 2004
Settlement agreement reached	September 2004

The NEPA protocol places citizens in a reactionary role: NIH and its contractors created the documents that framed the discussion. Some members of the public contended that their concerns were not addressed in the documents; when they raised these concerns in other venues, they were told that they would be addressed in subsequent versions. But when they finally received the draft EIS—a one-hundred-page document, which one person described as smaller than the Hamilton phone book—they didn't find answers, and voiced their complaints in the meeting: "Many people had questions they were told were going to be addressed in the EIS. You have an obligation to answer those questions. It's not an option." "I resent that people who are asking questions are labeled as afraid or ignorant. Maybe I won't be afraid if I get the answer." "Help me to not be so cynical about the process." People sought information in order to assess the risk the lab posed to their own well-being, and they expected those who prepared the EIS (an outside company) to do the work of collecting, analyzing, and evaluating the risk. They felt unable to quantify the risk themselves but retained the right to make decisions when presented with information they considered adequate. In actuality, the biological scenarios imagined by the public had never before been addressed through the NEPA process. The limitations of the protocols in recognizing evolving types of risk led to public disgruntlement about the process in general.

The range of impact of a BSL4 laboratory might be determined to be as small as the building's footprint or as vast as the entire nation or even the planet. To assert the regional or national benefits of the laboratory in the environmental assessment, its scope had to be broadened to also consider the potential harms on these scales.[69] The scope refers to the range of issues—actions, alternatives, and impacts—that the EIS addresses. Skeptics of the process claimed, "NIH has arbitrarily limited the scope of the DEIS. This is an obvious and transparent attempt to limit the scope to a location and budget that was predetermined to avoid considering a reasonable range of alternatives, and disclosing the rational[e] for the choice of location or budget tradeoffs."[70] NIH was in a complicated position. On the one hand, the laboratory had been described as critical to a nationwide security need and lauded as a regional center of excellence in the Northwest, but expanding the scope to a corresponding degree would require a tremendously comprehensive EIS. With connections to research projects on every continent and the importation of microbes from overseas, the scope of the work done in the RML could easily be called global.

Still, the various drafts of the EIS and the public comments affirmed clearly that the effects of a laboratory dealing with highly infectious diseases extend beyond the laboratory walls. One public concern, for example, was how the pathogens to be studied, such as Ebola virus, would be brought to the laboratory or moved between BSL4 facilities. The lab responded that pathogens were transported by commercial shipping companies that were regulated by the CDC. It contended that because RML did not oversee the training or certification of these agencies, this issue was outside the scope of the EIS. In alarm, citizens swapped jibes about shipments of Ebola being delivered by the FedEx truck, integrating the delivery company into their worst-case scenarios: What if the truck crashes? What happens if someone steals the truck, left running on the street while the driver pops into the drugstore?

Another concern centered on the resources available to the small town to protect its citizens if an accident occurred. The local hospital was not as well equipped as hospitals in large cities near other BSL4 labs, and people worried that should an incident occur it would be exponentially worse because of the limited capacity to respond. Thus the scope of the EIS was extended to include some community infrastructure in Hamilton and Missoula and the readiness of the community support personnel who would deal with an incident: a hospital in Missoula, for example, ultimately received money to build two isolation rooms.[71] Securing the community required people, training, and spaces that extended beyond the boundaries of the laboratory. Furthermore, by this conception, the evaluation of emergency situations—not just day-to-day practices—fell within the scope of the EIS. If the security of the facility were breached, the bounds of the laboratory would expand to include, at least on paper, the city of Missoula, sixty miles to the north.

QUANTIFYING LABORATORY RISK

"Risks only exist when there are decisions to be taken." Because Anthony Giddens's definition of risk depends on the position of the subject, in this case a "society increasingly preoccupied with the future (and also with safety)," risk is generated by people's actions, particularly decision making by entities that intervene on behalf of the individual or collective.[72] The EIS for the BSL4 expansion was a mechanism to transform the perceived hazards, dangers, and threats of the laboratory into a measured risk. The BSL4 laboratory was still an imagined future entity that could be eliminated through decisions in the present: the NIH could decide not to build, or construction could be prevented through legal actions based on the EIS. NEPA is employed to materialize a future outcome in a way that minimizes the risks to people and environments. Once those risks are calculated, governing agencies are assumed to be capable of making decisions that will mitigate risk. In Giddens's risk society, however, the unremitting advancement of science "manufactures" risk, producing a society in which the uncertainties of science cannot be dealt with except by further advances in science and technology.[73] These new risks have little historical precedent, and, as Petryna notes, "we often don't really know what the risks are, let alone how to calculate them accurately in terms of probability tables."[74] The expanding threats of bioterrorism made possible by the new biology can be countered only by the rapid growth of the science complex, which promises better technologies to facilitate a faster, more efficient response to any possible event.

A two-sided problem in identifying and assessing the risk of a BSL4 laboratory emerged through the EIS process: the preparers of the report did not include a full range of hazards, while community members put forth increasingly extravagant scenarios that resisted quantification. While imagined future catastrophe is routinely considered in environmental impact assessment, biological threats held the potential for incalculable risks. In its earliest version, the environmental impact statement stated that it is not possible to measure the risk of a BSL4 laboratory. The DEIS noted that the proposed action presented "remote increased risk to the community" but that the "potential added risk to the community from the Proposed Action cannot be effectively quantified."[75] Though the risk couldn't be measured, according to the DEIS, it could be mitigated through standard operating practices, for the "safety measures *inherent to* RML would effectively reduce threats of terrorism and the possibility of a release of a studied agent into the community."[76] Support for this claim cited the laboratory's past safety record without acknowledging the changed future conditions that created the risk, again failing to recognize that the idea of risk is entwined with the subject's perception of the future, not the past.

While some community members, including the mayor, accepted the risk as negligible and trusted that past record, others insisted that the difficulty of assessing risk was no reason not to try. The comments on the DEIS submitted by the three community advocacy groups argued that "the fact that it is difficult to assess risk in this case does not mean that it is impossible to quantify in an EIS." Further, they argued that even when the risk is minimal, it is still within the bounds of the EIS to study that risk, and asserted that "risk assessment is a common practice of the Federal Government."[77] Citizens ascribed authority over risk assessment to the federal contractor preparing the EIS and the groups who generate statistical information about risk, but they would not tolerate a finding of minimal risk. Through the authority of the EIS process, they sought affirmation of the "social risk" they experienced. The EIS also created a rationalized assessment of risk with the object of enabling intervention, a distinct type of risk that Mitchell Dean labels "governmental risk."[78] Assessment practices make risk knowable, but they also make it governable, binding the notion of risk with outcomes to be brought through governance.[79]

In asking for an assessment of risk, citizens demanded disclosure of the practices and procedures that created the risk. Through Freedom of Information Act requests, activists sought information about what microbes would be studied, the history of accidents at RML and existing BSL4 labs, and how Hamilton was chosen as the site for the lab. Furthermore, citizens placed the burden of proof on the agents of change. Gorman explained that the goal of their activism was to make the government accountable to the citizens: "We need you to prove you can do this safely and responsibly in Hamilton."[80] RML would be responsible for mitigating risk should the laboratory be built, and part of proving that they could do this effectively involved proving that they had clearly identified the risks. Still, some community members seemed unwilling to accept that the labs posed no threat, even when the NEPA process authoritatively illustrated this conclusion.

The final EIS incorporated risk assessment more fully into its methods, including sections titled "Community Risk" and "Risk Assessment Scenarios." The NIH-developed "maximum possible risk" model tried to build credibility by "simplifying assumptions [that] we know for certain are more unfavorable than any credible assumptions, . . . [building] extra confidence since the actual risks are certain to be less than the risks presented in the analysis."[81] The model worked through six "reasonably foreseeable, credible" threat scenarios, concluding in every case that the risk was "none." Statistics play an important role here, because some numbers were so small that they were rounded down to zero. One scenario concluded, "The calculated potential release described in this scenario would be 0.000011 spores. Since release of a partial spore is not feasible, this number is practically rounded to zero." Another asserted, "The risk of public harm is so minute that it may be considered zero."[82]

In the eyes of the public, however, minimal risk was not equal to no risk. Comments by one advocacy group argued that the assessment of risk was fundamentally different now because high-security science laboratories were more numerous than in the past. In other words, statistics based on the existence of two or three working BSL4 laboratories might be small, but more laboratories multiplied the risks. The FEIS still assessed risk based on precedent, a method that also proved unsatisfactory and even contradictory. A literature review of laboratory-acquired infections cited more than five thousand occupationally acquired infections since 1898, with six occurring since 1999 and one occurring in a BSL4 laboratory, then concluded that "the overall safety record of biomedical and microbiological laboratories also indicates that there is not a risk of accidental release."[83] Not only did the conclusion seem to contradict the evidence, but the reference to past events was not convincing in a future-looking climate of risk.

Finally, the unpredictability of human beings complicates the calculation of risk. "Scientists are so glued to their microscopes, they can't see the world around them. They're not infallible, even if they seem to be," commented Larry Campbell, a Hamilton resident.[84] The scientists who worked at RML lived in a small community where their foibles and fallibility were on display outside the laboratory, lifting the hem on that cloak of mystery that so often surrounds scientific work. People also did not have to look far beyond the anthrax incidents two years earlier to believe that some scientists may intend to do harm. Campbell recalled a meeting where someone suggested that a person with ill intent could load a cement truck full of explosives, drive it right down Third Street, and crash it into the laboratory building. In response, an NIH director said that the heat of such an explosion would kill the pathogens. "But people live through explosions—we see that all the time. Why couldn't a microbe? This dismissive type of analysis is totally dangerous."[85] The EIS proved unsatisfactory to residents because it did not provide a response to these worst-case scenarios, which resonated with new conceptions of vulnerability. One resident stated in the comment session on the draft EIS:

> And then it goes on to say that the "proposed added risk to the community from the proposed action cannot be effectively quantified"—and because it can't be effectively quantified, we'll just ignore it. And I think that is an egregious oversight. There's no discussion of the possible accidental or purposeful breach of security, the potential direct or indirect or cumulative effects of such a thing if it does happen. The "What if?" The "nightmare." . . . What happens if the worst happens? Because obviously it's a possibility because the whole document talks about taking, you know, reducing the risk. . . . But the risk never goes away. . . . What happens to the Bitterroot Valley if we do have the worst-case scenario? Let's have it out on paper. 'Cause it's a possibility. And don't ignore it. And don't ignore those of us that are concerned about it and treat us like we are stupid.[86]

Musings over hypothetical situations pitted community members against one another, creating a new politics marked, in Giddens's words, "by a push-and-pull between accusations of scaremongering on the one hand and of cover-ups on the other."[87] While Giddens seems to suggest that anyone from scientist to politician to layperson could proclaim the risk equally, citizens in Hamilton were not equally vested with authority to identify risk. Though some of their comments were addressed in the EIS, the document still dismissed most of the scenarios and information requests as outside its scope. Furthermore, some citizens were quick to label others as fearmongers. People who supported the laboratory were so convinced of its safety that they felt confident in accusing others of using scare tactics and lacking proof for their fears in the absence of any data. More important, the authoritative actions of the NIH, RML, and other government agencies were powerful enough to equate fear with ignorance.

If fear is assumed to result from ignorance, risk can be overcome through the production of knowledge and the education of the citizenry, demanding no material change in the laboratory. Petryna says that after the nuclear meltdown at Chernobyl, "life was perceived to be in the hands of an invisible all-knowing expert" who controlled the flow of information about health risks. In the face of a crisis, people sought information "to render an uncertain and unknowable world knowable and inhabitable in some way."[88] Knowledge of the physical world offered access to life and survival; communication offered access to that knowledge, but the state still controlled that information. The NEPA process exemplifies one way that the state assigns risk: by subjecting a new entity of unknown scope and impact to NEPA oversight, the governing agency retained the power to set parameters and quantify risk. Rather than presenting a full spectrum of risks to be evaluated and compared, the EIS dealt in absolutes.

Only some risks were given enough credibility to be quantified; all "negligible," "minimal," and "slight" risks were rounded down to "none."

Because of a limited production budget, the draft EIS presented to Hamilton residents in June 2003 considered only two alternatives: the "preferred action" and "no action." Of the 588 comments received during the scoping period, around 10 percent requested NIH to consider alternatives to building in Hamilton, particularly in the center of town. The stark choice put the focus on the potential economic benefits and precluded suggestions that risk could be reduced by considering alternative locations or situations, such as building the facility outside of town, downwind and downstream. If people wanted the lab to continue to enrich their community and support their nation's defense, their only choice was to accept the new laboratory as proposed. They could not claim or exercise their biological rights while still supporting the work and workers of the laboratory.

Advocacy groups contested that this approach violated the spirit of the NEPA process, which purports to consider a range of alternatives in order to select the least damaging option. An EIS is a document for assessment and is not to be used in "justifying decisions already made."[89] But RML had already begun contracting with builders, developing blueprints, and otherwise investing money in the project. It seemed that everything was in train to proceed, if only the EIS could assure residents that there was no risk of harm. Scientists were excited about the chance to be on the cutting edge of virology, and many saw this as a chance for Montana to play a role in the war on terror.[90] The post-9/11 state of emergency built public support for swift response and enabled national security actions to circumvent the due process of law. Citizen groups in Hamilton pleaded for time for the NEPA process to play out before the government acted. In this case, the perceived need to secure the nation by building a BSL4 laboratory drowned out the voices of citizens. In times when assessment of human rights is overruled by the apparatus of security, the state passes broad judgment on the worth of individual human lives.[91]

"THE MEMO"

During conversations with Hamilton residents in 2008, people repeatedly asked if I had seen "the memo," an unsigned document given to Alexandra Gorman in a stack of papers released in response to a Freedom of Information Act request.[92] Though no one seemed to have a copy of the memo, everyone remembered it, recounting the gist of the message in their own words: "Hamilton is disposable." "They chose us because we're away from major population centers." "The internal memo makes it seem like the worst is going to happen, so it might as well happen here." They presented this document as evidence that the selection of Hamilton was based on criteria other than the presence of RML and that deliberate decisions involving their own bodies had been made in selecting the site.

I found a copy of "the memo" on a website set up to protest the establishment of a BSL4 lab at Boston University.[93] Though it refers to a "clear and present danger posed by the daily threat of human and agricultural bioterrorism," it was written in December 2000, a year before the anthrax attacks raised the profile of biosecurity. The memo cites multiple reasons for expanding NIH facilities in Montana, including the availability of land on the large campus, collegial relationships between RML and Hamilton, and proximity to West Coast population centers. The argument that caught Hamilton's attention was: "Third, the RML campus is located in rural western Montana, well removed from major population centers. The location of the laboratory reduces the possibility that an accidental release of a Biosafety Level-4 organism would lead to a major public health disaster." Citizens in Hamilton clearly had a different conception of public health disaster, and they read in the memo an overall message that their lives were expendable.

These lines, though brief, framed a biosocial argument that when it comes to bioterrorism, the government does not value all citizens' lives equally and that those living near research facilities must assume risk on behalf of all citizens. When the memo was brought into circulation at a Coalition for a Safe Lab meeting, people began to imagine the geography of the Bitterroot as seen by the federal government. "For one thing, it would be really easy to quarantine the valley. There's only one paved artery in and out at both ends. At my most cynical, I can imagine them just shutting it off, then they could go back in and see where the bodies lay. In my own point of view, that kind of awful thinking is not unthinkable in the U.S. defense industry."[94] By describing the Bitterroot Valley as an acceptable sacrifice zone, the NIH devalued the lives of the people who lived there.

Several years later, people recited from the memo as an example of just how "ugly" things were during the BSL4 controversy. They still felt disbelief not just that the idea had been conceived but that the words had been recorded on paper, discrediting the well-crafted claims that the government meant no risk or harm to Hamilton residents. The memo pitted the people of this rural community against an agency that apparently valued bioterrorism research more than their lives. The David and Goliath archetype set small-town citizens against a national complex of science and government. While people didn't seem to believe their valley would deliberately be turned into a bioterrorism testing ground, the memo created the possibility that despite their activism, their small community might be denied their biological rights. In addition, the memo suggests the shift in the government's biosecurity approach that caused such upheaval in Hamilton, beginning with the announcement of plans to expand BSL4 facilities, was set in motion not by the events of September 11, 2001, but by a national agenda established prior to December 2000, which identified bioterrorism as a problem and established BSL4 research as part of the solution.

SECURITY SETTLEMENT

On September 24, 2004, Jim Olsen, Alex Gorman, and Larry Campbell sat in a chamber at the Great Falls courthouse negotiating a settlement with the National Institutes of Health regarding the future of the Rocky Mountain Laboratories. From 9:00 A.M. to 1:00 A.M., counsel shuffled between a room full of Hamilton residents and another packed with officials from NIH headquarters, setting out the terms that would end legal action brought by the Coalition for a Safe Lab, Women's Voices for the Earth, and Friends of the Bitterroot. These collectives had taken NIH to court to demand an EIS that took seriously the new forms of risk created by the biolab. The lawsuit demanded that NIH scrap the previous EIS and commission a new assessment by a new firm before proceeding with plans for the laboratory. Still feeling national pressure to build the BSL4 laboratory and on a schedule now held up by months of community protest, NIH offered to settle.

In the settlement, NIH agreed to a range of safety measures demanded by the advocacy groups. Gorman recalls feeling surprised that NIH agreed to everything they asked: "I should have asked for more, they were so agreeable." The terms of the settlement addressed risks the plaintiffs felt had not been identified in the FEIS, including training for local health personnel and the codevelopment of an emergency response plan with the county. Individual responses to the settlement were mixed. Some felt that their concerns over safety were being addressed, bringing the desired outcomes even though the protests had failed to lead to a satisfactory NEPA process. Others felt some disappointment that the issue did not go to court, believing that they "traded away an opportunity to hold NIH's feet to the fire to come up with a legally defensible EIS."[95]

By taking RML to court, activists found a way to insist on specific measures to protect citizen health and security. The EIS process had failed to delineate or address security concerns, but in court, NIH found a way to circumvent legislative mandates for impact assessment and meet the frantic deadlines of the national security complex by agreeing to certain tangible outcomes. Though the question of risk was not settled, certain practices would be undertaken to increase biosecurity. Neither the judge nor the EIS preparer nor the scientists bore the burden of proving that there was or was not a risk, but the NIH agreed to certain actions that citizens claimed would help them feel secure. The settlement was action-oriented, legally binding NIH to specific conduct regarding safety, security, and community involvement. Thus, biosecurity was removed from the realm of ideological assessment and translated into material practice. In choosing to settle, the community groups prioritized outcome over ideology, deciding that if the lab agreed to behave in certain ways, the quantification of risk was less important to citizens' well-being. In this case, as throughout the national security state, the actions to mitigate perceived risk were not necessarily informed by scientifically certain assessments of the biological threat.

Reflecting on these outcomes a few years after the settlement, while the IRF was under construction, Gorman suggested that the entire NEPA process and subsequent settlement would be a model for future BSL4 developments. Because of what happened in Hamilton, she believed NIH had to acknowledge that these laboratories have broad impacts and that citizens are attuned to microbial threats. Because of the community involvement, even to the point of the lawsuit, NIH "put a lot of systems in place they never would have done, and did a lot of additional planning and coordination, which they now realize was necessary to make this a more viable project."[96] The settlement also bound NIH and RML to the community, formalizing a casual relationship that had existed for generations and assigning the laboratory specific and significant legal responsibilities toward the people who live around it.

REMAKING RML FOR THE AGE OF BIOTERRORISM

As we walked three blocks from the lab to the Sunshine Diner, Marshall Bloom, the director of RML, talked about fishing and his advocacy for trout conservation in the Bitterroot River. A virologist by training,

Marshall Bloom worked at RML for thirty years before becoming director in 2002, in the midst of the controversy over the lab expansion. Six years later, he seemed relieved those heated debates were behind him, talking about the research he would begin in the new laboratory when he left the directorship. As he explained the systems of redundancy and security in the BSL4 laboratory, which he had clearly recited hundreds of times, it was evident he believed in the work of RML and the importance of studying infectious diseases in a world where microbes can still cause social upheaval akin to what the Bitterroot experienced a century ago.

While he acknowledged that the lab was built with bioterrorism and biosecurity funding, he said, "We don't consider ourselves as working on bioterrorism, but as working on emerging infectious diseases."[97] Still, the laboratory Bloom built is a product of bioterrorism, and it looks different from all other labs because of the community's multiyear protest. The material changes to the laboratory brought by the contestation in Hamilton show that the manipulation of space and the endless production of knowledge are primary means for convincing citizens that the state is caring for their biological survival.

Like most high-security labs, the laboratory in Hamilton has submarine doors, four-foot-thick walls coated with epoxy (because air can move through cement), and an elaborate system of ducts and vents to continuously draw air into the laboratory and purify it on the way out. An upper story houses the water and air purification systems. The space is designed to *contain,* to secure microorganisms that are transported through air, water, and flesh. Autoclaves are the entrances and exits to the biolab, sterilizing instruments and even the carcasses of animal subjects as they leave the laboratory. Human bodies that move out of the secure space must pass through a decontamination process that transforms them from potential carriers of lethal disease into neutral entities. The ritual of bodily inspection has been technologized from the early tick lab where, "on going out of the department, the men in charge of tick rearing leave their white coveralls to be baked in an electric oven, take a shower bath, and examine their bodies carefully for ticks before a three-paneled mirror."[98]

If the cleaning sequence is violated or fails, whether intentionally, inadvertently, or in an emergency, the departing body exposes the community to risk. Through needle pricks or inhalation, lab workers can also take microorganisms into their bodies, beyond the reach of the decontamination process. The settlement determined that the transportation of infected bodies would also be the purview of RML. It further stipulated that NIH would evaluate and regularly assess the function of the lab's air and water systems and that it would comply with federal regulations in transporting BSL4 agents. By determining that RML has jurisdiction over its water and air emissions, and by recognizing the lab's partial responsibility for shipping, the settlement codified three material outputs from the lab, challenging the assumption that a BSL4 facility is totally secured and insisting that its outputs should also be assessed as part of the lab space.

To minimize fear, the laboratory must appear secure to the public. After the 1927 court case against the lab, RML decided to build a moat around the laboratory building, reinforcing the folk wisdom that ticks could not cross the Bitterroot River by constructing a water barrier around the laboratory itself. Floors and walls were built without cracks and crevices, an advanced technology in the day, and tick holding rooms were designed with rounded corners to alleviate the fear of ticks lurking in tight, dark spaces. These physical alterations to the building materialized the idea that the laboratory was a secure, safe space and signaled to the community that the lab was concerned with protecting their health and well-being. These material gestures communicated concern with citizen safety without having to provide further evidence of safety precautions.

The deployment of such visible security measures raises the question of whether institutionalized security measures that are less visible (such as air filtration systems) can by themselves sustain community trust in the laboratory. In the IRF, many of the security measures are internal and inaccessible to inspection by the community. To make these technologies visible, the NIH settlement required disclosure of status reports. These numerical reports require expertise to decode the data, and therefore still require the public to trust the agents who run tests and report results, but they are today's equivalent of the moat, an intervention formulated to match the cultural fear of microbial risk.

The citizen protest of the IRF expansion also institutionalized ways of communicating risk, a change needed because the physical closing of the laboratory to the local community required new ways to make the laboratory space knowable, familiar, and therefore less frightening. The lab hired a public affairs officer to develop and present a unified message to the public. The settlement also stipulated that the Community Liaison Group would continue to meet "at a location outside the RML campus and . . . allow a reasonable time for public comment at such meetings."[99] NIH further promised to communicate information about the pathogens being studied, their medical symptoms, accident reports, and safety inspections to the public and their representatives. Thus the settlement ratified the perception that biosecurity hinges on strong intergroup communication, binding local entities to practices happening inside the laboratory walls. The settlement agreement also stipulated that community members would be represented on RML's institutional biosafety committee (IBC), the group that reviews proposals for research. Typically such boards are internal to institutions, and all members are affiliated scientists and officials. Gorman, who was one of the first nonscientists on the board, recalls a lot of tension during the first meeting, while long-standing committee members figured out what the three public representatives had to offer. Bringing local community members onto this board was an innovation, one that acknowledged that nonscientists could contribute to conversations that center on the ethics, outcomes, and methods of complex scientific work and might be entitled to a voice in stipulating the conditions under which it is done.

The assertion that knowledge conquers fear was reiterated by scientists and citizens alike throughout the BSL4 contestation. Advocacy groups settled with the NIH subject to the provision that NIH would maintain communication with the community. The biosecurity created through the manipulation of physical space is sustained by the production of knowledge. When citizens cannot enter the laboratory space, they can know that space only by the information that flows from it. In Hamilton, trust was developed through the long-standing and economically beneficial presence of RML in the community, but it had to be reconfigured in a world where national security concerns seemed to override citizens' own biological security.

These efforts to engage the community in the governance of risk show the first steps in building trust and transparency between citizens and government.[100] Following the outcry and hyperbolic imaginings of the BSL4 debate, risk governance may be finding its way in Hamilton, Montana. The question remains, however, whether risk can be mitigated through improved forms of governance. Perhaps, indeed, this is fundamentally a question of perception management: in Hamilton, the right of community members to speak out through the CLG and IBC seems to provide a sense of security to its citizens. Whether this is an effective form of governance can be tested only as the governing agency, in this case NIH and RML, transforms the public knowledge created in these venues into institutional practices and policies, bringing measurable outcomes as a result of public engagement.

"I NEVER WANT TO SAY, 'I TOLD YOU SO'"

A shiny new building sits today on the RML campus, reflecting the snow-capped peaks of the Bitterroot Range in its shatterproof windowpanes. Fishing season is in full swing, and residents are heading out to enjoy the brief Montana summer. Judging by their conversations, they seem more concerned about drought and forest fires than Ebola and smallpox. People have not forgotten the controversy over the lab, but they seem to accept its presence. Mary Wulff still lives in Hamilton, though she is skeptical about the future: "I never want to say, 'I told you so.' If I do, we're all toast." She believes that the collectivities that formed around the laboratory protest will endure, playing out in new ways in the future.

The present lack of concern among residents raises the question of the duration of biosociality: do citizens unite to advocate for their biology and then disperse when a threat no longer exists? Certainly a temporary biosociality mirrors the dynamic quality of life itself, for how could subjects build static identities on terms in such flux? The outcomes in Hamilton, however, suggest that biosociality gains potency from the shared vulnerabilities created by modern natures. Perhaps the biosocial communities dissolve not because of material changes in the environment—for the biological risk in Hamilton endures today—but because they become less useful as political tools. When biological claims become less effective in accessing social and economic inclusion, they also become weaker as social ties.

The rise of the security culture in the United States has remade the terms of citizenship. Petryna claims that "citizenship is now charged with the superadded burden of survival," rather than being founded on tenets of civic participation and human rights, because those principles cannot guarantee basic survival.[101] If survival is the primary biopolitical goal, any threat to survival must be handled before and in conjunction with all other negotiations of citizenship. If the state is parceling out the intimate care of citizens to individuals and corporations, as Rose suggests, perhaps it is because the burden of watching over citizen survival has grown with the rise of the national security state.[102] When the issue of survival overwhelms negotiations of citizenship between state and subject, less effort can be directed to the discussions of individual freedom, civil rights, and intimate care. Bioterrorism has effectively moved questions of collective and individual survival to the fore of political interactions.

The story of a twice-contested laboratory in rural Montana illuminates how citizens claim biological rights based on changing cultural knowledge of science and terror. Motives of fear, trust, risk, and prosperity worked in conjunction through the hundred-year project of making Rocky Mountain Laboratories, simultaneously shaping the surrounding community. Through all these negotiations, community members have often felt helpless in the face of the power and politics of the science regime. A witness in the 1927 trial explained why his neighbor refused to join the case against the labs: "He is very much in fear of ticks, and does not feel there is a chance to get rid of this laboratory. He thinks the Government are too powerful, and it was useless to contest the Government." Though the citizen collectives that came together at various times to protest Rocky Mountain Labs did not stop the construction of the laboratories, they exemplify the desire of citizens to act on behalf of a biological present. The protest of the BSL4 expansion tested new conceptions of the world and its risks alongside the rise of modern microbiology and an endless war on terror. These "serious speech acts" of citizens and scientists have "problematized" the modern laboratory space, bringing objects into "the play of true and false." As Rabinow explains, "The reason why problematizations are problematic, not surprisingly, is that something prior 'must have happened to introduce uncertainty, a loss of familiarity; that loss, that uncertainty is the result of difficulties in our previous way of understanding, acting, relating.'"[103] The new biology creates countless uncertainties in the human condition, as do the politics of the war on terror. In Hamilton, a familiar space received new scrutiny because political responses to bioterrorism threats introduced uncertainty into the local landscape, rendering the lab unfamiliar and leading to renegotiations of fear, harm, and risk. Bioterrorism has material effects because the work to redefine the terms of citizenship brings about new spaces and remakes existing spaces, such as Rocky Mountain Laboratories, according to changing perceptions of biological security.

4

Agents of Care

Bioterrorism Preparedness at the CDC

On the second floor of a glassy twelve-story building in the Atlanta suburbs, a room hums with up-to-the-minute technology, real-time surveillance maps, and secure telephone lines to the White House. On most days, this room sits empty, except for two workers staffing a twenty-four-hour hotline for public health workers. During public health events, however, the rows of workstations and high-tech meeting rooms crowd with managers, scientists, logistics coordinators, and media personnel from the Centers for Disease Control and Prevention (CDC). When the command center fills, whether during a health emergency or during a simulation exercise, CDC becomes a site for mobilizing swift government action to protect lives. This is a new role for an agency that built its reputation as the premier source of expertise on all things related to disease.

The CDC created the Emergency Operations Center (EOC) in 2005 with counterterrorism funding and private donations. Colorful pennants on the walls commemorate all the occasions the EOC has mobilized, for events ranging from bioterrorism training exercises to the disintegration of the space shuttle *Columbia*. Every incident, large or small, real or simulated, is directed from this central command space. The EOC showcases a new variety of expertise on crisis management, work that structures the systems of authority during health emergencies. When people come to the EOC to command, calculate, communicate, and act, they legitimate the work of government to manage citizens at risk. This chapter explores how redirecting federal funds to the CDC for bioterrorism preparedness established the agency as the nucleus of the nation's biosecurity program. The influx of funds drove the agency to modify its expertise, its mission, and its buildings to align with new ideas about biological nature created after 9/11. In turn, the CDC has become a primary site for producing new microbial natures through practices centered, more than ever, on preparedness for biological events. Evidence from the first decade after the anthrax attacks of 2001 shows that CDC remade its science, surveillance, and public communication practices in response to national fears of bioterrorism. Through the CDC, the federal government has created new spaces, such as the EOC, and new materialities, such as the Strategic National Stockpile of medical countermeasures, to manage these fears. Such measures have nudged the locus of public health practice away from local communities and toward the federal government, meeting the aims of the national security state to consolidate power over citizens' biological lives. The public health practices institutionalized by the CDC constitute microbes as threats to collective life and bring scientific expertise to bear on the management of human and nonhuman natures.

Consistently ranked among the most trustworthy government agencies, CDC has institutionalized particular ways of managing health and disease throughout the world.[1] The agency brokers the exchange of expertise between laboratory science and public health practice. Historically, CDC has claimed to offer its expertise only at the behest of other agencies, primarily by sending field agents to advise local governments and health workers during epidemic events. Today, although CDC can still deploy experts in the field within hours, the EOC enables CDC to direct the disease response from its national headquarters by means of communications technologies. Ironically, although the command center was created to bring experts and political actors into a common space, the agency is now implementing ways for people to report to the EOC from remote locations (including their homes), acknowledging that the bodies of CDC employees may carry biological risk and that physical proximity may spread disease. EOC protocols require individuals to stand ten feet apart during pandemic events, and dispensers of hand sanitizer mounted in every room remind employees of their own vulnerability to disease. Mechanisms of security govern this space and the bodies that work here. As CDC experiments with its new command center, it builds expertise that extends beyond laboratory studies of microbes into the governance of human life.

With the rise of a modern security state from a war on terror, the public health system in the United States is being militarized as never before. Countless daily acts of care have been harnessed by security systems as displays of government intervention on behalf of citizens' health. In the name of national security, the government has created, financed, and maintained the global health work of disease eradication, scientific study of pathogens, and highly securitized biological laboratories. These institutions create a coherent national identity by generating a type of citizenship centered on health risk and the protection of human life. They reach to the biological core of society, creating categories of life and death, illness and wellness. Because they extend so far and because they are naturalized as part of the human biological experience, public health systems provide a potent tool for managing the risks of modern living.

The coexistence and co-constitution of microbes and humans inevitably create an environmental politics around cross-species interactions. Some of these interactions are also understood to bring harm to human life, both individually and collectively. Thus, the public health apparatus exists to care for citizens by managing risks in the environment and in their own bodies. The belief that nature could be managed for the health of populations rationalized governance and established the care of people, whether through sanitation, vaccination, or education, as an appropriate and expected responsibility of modern government.

Systems of care traverse private and public life. Acts to guard individual health also work as acts of nation building. A vaccination, for example, is an intimate, individual act that changes the human relationship with microbial nature, but when repeated by many individuals, mandated by governments, or coerced through peer pressure, vaccinations have a broad social effect. Vaccination is both an act to protect individual health and an act to protect the community and nation. Bioterrorism preparedness, smallpox eradication, the war on AIDS, food-safety recalls, and the Affordable Care Act are all part of the national experience of health in the United States.

When people experience health collectively, Foucault argues, they can be known as a population.[2] At the level of a population, some deviations in health can be deemed statistically normal and therefore acceptable. Other abnormalities fall into the realm of the unacceptable and must be policed. This statistical accounting of the health of a population delineates acceptable levels of disease and risk, which governments must maintain. Through its data collection and surveillance expertise, CDC has become the lead agency in setting parameters for acceptable U.S. deviance in the population, sending alarms when there are abnormal rates of anything from *E. coli* outbreaks to obesity. Care thus requires continual surveillance and diligence on the part of the caregiver to act and react to entities that threaten the system.

Liberal governments were created to care for citizens and to ensure their human rights. Public health systems coalesced in the twentieth century around the ideal that all humans could be made biologically well through the strategic use of technologies and the authoritative hand of government. As a bearer of deviance and a holistically unpleasant experience, disease presented an enemy for social war against nature. People "fought" germs and pathogens through municipal sewers and septic systems and through the diligent use of cleaning products within the home.[3] Political health reforms were a show of governance that united people by overcoming a common vulnerability while garnering support for the extension of government rule into people's homes and bodies.

Bioterrorism has extended the role of government in the care for citizens' bodies, directing attention beyond known microbial threats to future possibilities. The physical experience of biological attack does not enter into this calculation, because people preparing for bioterrorism may be entirely healthy, though they carry future biological risk. Bioterrorism preparedness activities work to keep people safe in the future, not just to care for their health in the present. Future natures remain unknowable, except through predictive science, which transforms the uncertainty of future disease into calculations of death and social demise (see chapter 5). The technoscientific transformation of microbes into weapons of mass destruction shifts the parameters by which disease is known, requiring new equations of biological risk and new interventions to preserve the nation's health. The politics of bioterrorism center on the management of populations to prevent biological attack and assure future life. Through the national attention to bioterrorism, including massive government allocation of resources, new public health systems have taken shape, and the CDC, the nation's premier institution on public health, has been remade as an expert agency on national security. In turn, bioterrorism has brought health-care practices into the service of the domestic security agenda.

Because of CDC's scientific expertise on biological agents, it was brought under the umbrella of emergency response. The agency has been ideologically and materially restructured around emerging notions of biosecurity. The changes at CDC in the early twenty-first century show that the traditionally local focus of public health has shifted to the national level in the name of producing a timely, nationwide response to disease threats. These shifts have also led to changes in the ways citizens' bodies are scrutinized and managed for risk and how care is delivered. Disease control practices deemed necessary to contain an epidemic may violate citizens' privacy and autonomy, particularly when the population is understood to be continuously at risk of biological attack. Yet if, as one team of experts has claimed, merely preventing a biological attack is "not sufficient" to achieve biosecurity, then the objectives may never be attained, leading to a state of exception characterized by a never-ending threat and demanding perpetual vigilance on the part of the government and all citizens.[4]

To see how the war on bioterror has shaped the public health system, I went to CDC during the fall of 2009. I hoped to understand how modern microbiology entwines with prominent cultural fears of bioterrorism to bring about specific outcomes in public health practice and people's experience of disease. Bioterrorism preparedness remade scientific practice at CDC through the reorganization of laboratories and reallocation of resources. Interviews with scientists and managers showed that the biopolitical response to terrorism ranges from keen attention to workplace behaviors to the creation of new spaces and systems that enable a large-scale emergency response. My findings, primarily derived from interviews conducted with officials at CDC around that time, show a clear militarization of care. Public health looks different when recast as biosecurity, for the public acts within this security climate to create the institutions of care they desire. This intimate work to create human natures is the political praxis that constitutes our morality and justice.

THE ALLIANCE OF WAR AND PUBLIC HEALTH

Fueled by billions of dollars in bioterrorism preparedness funding, efforts to prepare the health system for biological attacks have been under way in the United States for more than a decade, bringing a new level of militarism to the everyday practices of health and wellness. As the federal agency charged with "controlling the introduction and spread of infectious diseases," CDC's work has structured bioterrorism preparedness efforts at all levels of public health, from the White House to the community health center.[5] CDC agents were on the front lines for the anthrax events of 2001, and soon afterward, the organization's director, Dr. Julie Gerberding, pledged to Congress that CDC would be a part of the domestic terrorism response: "The events of September and October 2001 made it very clear that terrorism is a serious threat to our Nation and the world. . . . CDC has made terrorism preparedness and emergency response one of two overarching agency goals and has built an infrastructure to catalyze and implement biodefense activities."[6] In 2002, the agency established the Coordinating Office for Terrorism Preparedness and Emergency Response, and by 2012 the office managed $1.3 billion appropriated by Congress for terrorism preparedness, about 23 percent of CDC's $5.7 billion budget.[7]

Bioterrorism infused CDC with a sense of urgency, expanding the agency's nominal focus on disease control and prevention to include emergency *response*. CDC declared in 2006 that it would increasingly focus on using scientific research and practice scenarios to improve emergency response.[8] Again, Gerberding identified terrorism as the catalyst for such a reformation of national health strategy:

> The philosophy of public health during the 20th century has been to prevent natural outbreaks. In the 21st century, however this is not enough. The threat of terrorism necessitates that we improve our public health and medical systems so that we can respond with greater flexibility, speed, and capacity to handle mass casualties and large-scale emergency response in coordination with our traditional emergency response partners as well as those at Department of Homeland Security (DHS) and Department of Defense (DoD).[9]

Gerberding explained CDC's alliance with the nation's military agencies as part of a new regime of disease control, in which biological agents that have been manipulated by human technologies demand the expertise of scientists as well as soldiers. By describing microbes as a threat to national security, the U.S. government enlisted its own public health system in the war against terrorism. Moreover, because microbial infection is battled at the intimate level of individual bodies, the security state effectively recruited every biological citizen to the battle.

New biological technologies and new understandings of microscopic life are similarly extending scholarly theorizations of Foucault's biopolitics to consider how microscopic life shapes society through microbiopolitical practices.[10] The nonhuman natures of the human body enter the social system through health and medicine, making these into sites where the biological clearly becomes political. Tracing the historical trajectory of medicine in the United States, Ed Cohen argues that

> medico-political responses to epidemics . . . reveal the ways that medicine increasingly legitimates its protocols in terms of political values and outcomes to extend its own power and expertise. Through this hybrid process, medicine begins to supplant religion as an authoritative basis for making political decisions about the public good.[11]

As health accrues social capital, medical practitioners and scientists gain power through their control of the systems of knowledge production. As the political value of security rises, public health specialists bring authority and expertise to the realm of biological security, thereby sustaining their relevance and, as Cohen argues, legitimating health expertise as the basis for political actions. Considering the divergent risks and interventions enabled by CDC's involvement in national security elucidates how bringing the biological world into the global war on terror has allowed governments to access citizens' bodies in new and distinct ways.

Though presented by Gerberding as a new association, the alliance of practitioners of health and warfare characterized military conflict throughout the twentieth century. The germ theory of disease brought infection into the calculation of security by identifying a source of disease outside the human body. Medical surveying, federally instituted during World War II with the establishment of the Medical Survey Program to track the health of Selective Service registrants, recorded information about a population's health, including good health as well as pathology. Mechanisms were developed to measure and track health in terms of a national population.[12] Global warfare and the acts of nation building that characterized the modern age also exploited emerging fears of external, transmissible, tropical, parasitic, and endemic disease. Soldiers fighting abroad encountered new microscopic enemies, and discourses of conquest began to form around global disease eradication. On an unprecedented scale, public health practices that might be deemed intolerable infringements on civil liberties in peacetime were sanctioned by the scientific promise that disease could be eliminated through the careful management of populations. By the time the ideal of a disease-free world faded in the late-twentieth century, disease-control mechanisms were entrenched in practices of warfare, citizenship, and governance.

Foucault theorized a changing relationship between people and disease based on the ability to calculate risk and create a population, a collection of living beings defined by their biological and pathological characteristics and subject to common governance. Through the mechanisms of security, an acceptable level of disease within a population can be calculated, acknowledging morbidity as a part of normal, healthy life while still enabling the security apparatus to work against deviant cases and allowing individuals to act on their personal will to be healthy.[13] The population encompasses both those who are ill and those who are not, and disease is part of the risk that defines the population. Thus the healthy population is governed by the possibility that disease could be introduced, unknown to the individual or the state, at any moment.

Gerberding's call for flexible and fast response to a disease emergency was a reaction to an emerging paradigm of disease in which science and technology expand the potential for humans to use germs as weapons to cause harm. While CDC scientists have expertise on how microbes live, grow, and move within populations, weaponization breaks these bounds and creates unknowns, diminishing the expectation that disease can be controlled. By creating the possibility of a seemingly limitless number of biological threats, bioterrorism changes what it means to be "prepared" for disease. No longer can individuals take directed actions to prevent a finite number of risky diseases, for disease could come from anywhere, including new strains concocted in laboratories: it can no longer be known, as Foucault explains, "in terms of the calculus of probabilities."[14] With less ability to act in the interests of their own health, individuals seek new knowledge and expertise, leading to an expansion in the knowledge-producing role of government. This new understanding of disease has as much to do with the militant protection of national borders and the policing of citizens as with the scientific study of microbes.

Aligning terrorism with disease allows social controls long used to secure the population against germs to be reproduced in national security, enlisting all subjects in the effort to determine and maintain a "normal" condition of the population. Because contagion builds fear of human interactions by transforming all bodies into vectors of disease, healthy individuals govern themselves out of fear of contracting disease, as evidenced by people who donned surgical masks in public during outbreaks of SARS in 2003 and H1N1 flu in 2009. Minimally disruptive techniques like vaccination and social distancing aim to interrupt disease transmission while relieving the state of the responsibility to enforce a quarantine. The use of such biopolitical technologies extends to apparently healthy members of a population who are potential carriers of disease. All citizens are enlisted as agents of biosecurity.

Disease preparedness differs from prevention in its focus on infrastructure rather than a particular disease, and in the anticipation of future threats rather than the analysis of past events. Preparing for a disease event anticipates that the health status of the population may change, likely dramatically. Andrew Lakoff and Stephen Collier argue that preparedness practices create vulnerability by producing knowledge of a threat that does not yet exist.[15] CDC's bioterrorism preparedness activities in the first decade of the twenty-first century changed public health infrastructure and practices even in the absence of a large-scale biological event, with three particular outcomes: the normalizing of the fear of bioterrorism, the multiplication of biological risk, and the construction of new health systems.

If the objective of public health is to maintain the status quo, then the focus on bioterrorism has potential to normalize social fears along with existing health problems. Lakoff and Collier call this the "emergency modality of intervention," an emergency management strategy that "does not involve long-term intervention into the social and economic determinants of disease. Rather, it emphasizes practices such as rapid medical response, standardized protocols, . . . surveillance and reporting systems, or simple technological fixes like mosquito nets or drugs."[16] Though CDC is involved in a range of long-term health-promotion initiatives, such as antismoking and antiobesity campaigns, the agency's 2007 budget itemized $1.6 billion for infectious diseases and an equal amount for terrorism, the largest line items aside from vaccines for children; in contrast, health promotion received $222 million. This funding favors planning a bioterrorism response over inquiring into the causes and outcomes of disease.

Second, preparedness practices multiply risks: the population is perpetually threatened by health catastrophes ranging from bioweapons and pandemic disease to flooding, hurricanes, and pollution. Not only does the endless threat justify an armed and ready health militia, but when CDC responds to disasters like hurricanes or earthquakes, the militarized techniques of emergency response become part of the standard for disease control.

Third, CDC scientists frequently suggest that preparedness technologies can bring much-needed upgrades to public health infrastructure. Pamela Diaz, associate director of science for CDC's Division of Bioterrorism Preparedness and Response, asserted that "bioterrorism funds helped move communicable disease investigation into the twentieth century. It did, and we were not going to get that money any other way."[17] The preparedness infrastructure, however, looks different from the prevention infrastructure, and the question remains of how the new equipment, staffing structure, and research priorities resulting from bioterrorism funding shape public health work. Whether it is diverting attention from more-critical disease concerns or beneficial to the public health system overall, bioterrorism preparedness builds a system that sustains the belief that a terrorist event will happen and that the government will intervene through local public-health agencies.

The following discussion considers how three areas of CDC's work—the scientific study of disease and human disease behavior, surveillance of the population, and communicating preparedness interventions to individuals—have changed in the preparedness environment, creating a security mechanism that regulates social interactions at the national level while producing citizens who work daily to secure the nation against the bioterror threat.

SCIENTIFIC STUDY

"Remember, at CDC, data is king! It's the lifeblood of the organization, the mother's milk."[18] This CDC scientist described science as the leviathan driving CDC actions, determining what CDC does, how it intervenes, and how it spends money. At CDC, the course of science is also directed by a strong belief that the work has application in public health practice. Dixie Snider, a former chief science officer, explained that in CDC laboratories "there's this idea that there's some practical issue to be addressed. We don't have the orientation . . . that scientific knowledge is intriguing and interesting on its own merits, and if there's a practical spin-off that's great. As a public health agency, we don't have the dollars for that kind of research."[19] At CDC, science is a tool in the service of public health practice. Snider believes organizational changes reflect a shift toward broader thinking about disease and environments. Research divisions once structured around the study of specific diseases have expanded, and programs like the preparedness division apply science to questions of risk assessment and population management to inform public health practice as it relates to many diseases.

In 2006, CDC published *Advancing the Nation's Health: A Guide to Public Health Research Needs, 2006–2015*, a 139-page compendium outlining the agency's research priorities as established through internal working groups, input from employees and partner agencies, and formal public comment. The guide claims to identify gaps in knowledge whose elimination will justify public health interventions and improve the effectiveness of public health work. Influenza is the only disease given its own research category. The other research priorities follow the drift toward preparedness, and preparedness itself is the second of seven research areas, for "although emergency public health has always been a public health activity, health services research in preparedness and response must be made a priority."[20] CDC is not alone in its focus on preparedness. An inventory of government-funded research on national health security from 2003 to 2010 found that 78 percent of studies focused on preparedness as opposed to mitigation, response, or recovery.[21] The volume of research on preparedness reflects the uncertainty even experts feel about the future threat and the desire to garner knowledge that will make deterrence possible.

CDC's "examples of priority research" for preparedness focus on people and infrastructure rather than microbes, such as the goal to develop "reliable, valid tools and strategies to profile the vulnerability of communities along multiple sociocultural and community dimensions, including the mechanisms responsible for health disparities."[22] This agenda assumes not only that communities are vulnerable but also that scientists can quantify and measure the factors that produce vulnerability and use these data to enhance public preparedness. Because the research imagines communities as groups of individuals with predictable behaviors and characteristics that produce vulnerability—a Foucauldian population—these studies facilitate health interventions motivated by a quantified assessment of people and risk. Such research generates expertise on human behavior, not on disease and microorganisms, and this distinction has the potential to change the role CDC plays in responding to crises. No longer solely an expert on microbes, CDC will respond to an event (or even anticipate an event) with tools for profiling sociocultural vulnerabilities and the mechanisms that create differences within communities. Such research holds high potential to inscribe the social conditions of a population, including categories of race, class, or gender, as determinants of biological risk, backed by expert disease science. One need look no farther than the HIV/AIDS crisis in the 1980s to see how profiles of vulnerability and risk cohere around categories of people—in that case gay men, particularly gay men of color, as well as the urban poor and intravenous drug users.[23]

Scientific research also has specific application in setting standards that define biological security. For example, the anthrax events of 2001 required CDC field scientists to determine when people were safe from the microbe. When workers were cleaning spaces contaminated with anthrax spores, they confronted the question "How clean is 'clean'?" CDC had to quantify and prove cleanliness, a task the agency continues to debate as it develops standards for fieldwork. To prove that health workers have assessed a biological hazard and mitigated the risk, they must quantify the security or cleanliness of the affected space. Diaz explains that achievable objectives supported by scientific research may still not alleviate public concerns: "How clean does it have to be? How do you translate the risk of a spore or two spores in light of a human being? How accurate does a test have to be?"[24] CDC is directing preparedness funds at answering these questions, quantifying risk in terms of individuals and populations. Significant amounts of time and resources may be spent in pursuing scientific evidence to "solve" the crisis and assure policy makers and the public that they are secure, even as the methods of science employ different standards of certainty. Efforts to develop better field assays, establish protocols for handling potential bioterrorist agents, and collaborate with national security agencies generalize the public health response and build CDC's expertise on bioterrorist events. This expertise drives further research, though one director in the bioterrorism preparedness office claimed that in his experience, public health decisions are based upon evidence, while "bioterror decisions are based upon politics."[25] While all science is political, the attention to and demand for timely answers to bioterrorism may impede the scientists from formulating pertinent questions.

A particular challenge in preparedness research is the rarity of emergencies, which limits the amount of field data that can be gathered and analyzed. Because the intentional release of biological agents changes how microbes move within the population, scientists must use existing knowledge of microbial nature to imagine how disease *might* spread if used as a weapon. To compensate for the lack of field data, scientists are increasingly turning to drills and simulations, such as those conducted at the EOC, to imagine the outcomes of bioterrorism events. In 2013, the EOC was "activated" for 65 events and 135 exercises.[26] Through these exercises, CDC employees supplied data for preparedness research, working toward a goal of ensuring "scientific rigor in the design, implementation, and evaluation of drills and exercises."[27]

As a scientific method, such simulations fall somewhere between observation and experiment and are notably different from the laboratory science and fieldwork that built CDC's reputation. Scenarios aim to predict possible human responses and event outcomes based on laboratory knowledge of disease contagion and assessments of vulnerability produced by public health research. The slippage between processing data and producing information pervades simulation and scenario enactment. Scientists who think through the potential problems of disease using simulations materialize futures that enable actions in the present.[28] It is critical that the limitations of simulation-based research are evaluated and clearly communicated to science consumers and policy makers.

Access to bioterror preparedness funds has enabled CDC to establish expertise not just on disease but also on human behavior. Its disease-surveillance practices map the population as a network of people and places made vulnerable by lines of interaction and weaknesses in infrastructure. Scientific study of this data creates new ways of knowing human behavior. Priscilla Wald has argued that disease science has "harnessed the authority of science to depict the medical implications of the changing spaces, interactions, and relationships attendant on urbanization and industrialization."[29] By explaining the world in terms of disease, scientists reifiy the biological impacts of social interactions, subjecting communal behavior to the rule of health experts. As it grows into its new role as a bioterrorism authority, CDC brings these social networks under the scrutiny of national security concerns, using science to rationalize security actions.

SURVEILLANCE

Like the famous horseshoe seating of NASA's mission control center, all workstations in CDC's Emergency Operations Center are oriented toward a large bank of screens covering the long wall of the center room. On the day I visited, a screen on the far right showed the day's scheduled events, while screens on the far left streamed CNN and Atlanta news. In the center, a half-dozen screens displayed health-related information from around the world. One U.S. map showed the number of flu outbreaks in the country to date; another map charted salmonella outbreaks in North America. A global outlook map tracked diseases of interest, as reported over the previous two weeks, and another showed a satellite weather image tracking the advance of Hurricane Paloma. One screen listed twenty-three individuals on the airline industry's "do not board" list, primarily individuals with tuberculosis.

This electronic wall created an image of the nation's health, an up-to-the-minute snapshot of the people and places being monitored by the health sector, situated between the day-to-day operations of CDC and the events presented by the news media. The effective collection and distribution of public-health surveillance data are at the core of CDC's public health mission. The gathering and use of surveillance data also play a role producing a population that is vulnerable to bioterrorism and therefore in need of securing. As CDC's surveillance capabilities have expanded with bioterrorism funding, its stores of national health information have increased.

FIGURE 14. Emergency Operations Center at CDC. Photo: Centers for Disease Control and Prevention.

Public health surveillance involves the systematic collection of data on disease and health behaviors and the analysis of that information to direct health practice. Surveillance methods include laws that require health care providers to report certain diseases to local health departments as well as voluntary surveys that solicit information about health practices directly from the population. Information can also be gleaned from vital records, hospital reports, laboratories, and police reports. William H. Foege, a former director of the CDC, describes surveillance as the basis for all disease-control programs: "One could not work at CDC without internalizing the idea that disease control requires accurate knowledge about the disease and its environment and that this knowledge is obtained through surveillance systems. Response, or control, was based on surveillance findings."[30] Surveillance work binds microbes, humans, and landscapes in datasets that give material form to disease and serve as the basis for management actions.

New technologies are changing traditional health-surveillance techniques. For example, CDC is working with Google to explore ways of tracking Internet searches for disease symptoms in order to locate outbreaks in their earliest phases, before individuals start going to the doctor. Environmental sensor technology, as used in the Department of Homeland Security's BioWatch program, attempts to locate disease-causing microbes in the air before they find human hosts. Though these technologies fail to achieve the accuracy rates required of singular diagnostic tools, researchers continue to refine them and promote their use alongside more traditional individual diagnosis. Ultimately, disease surveillance targets microbes, but humans are implicated as the microbial hosts.[31] When it comes to surveillance, the microbe and human are inseparable, for disease is measured as the distribution of affected individuals in a population at a particular moment in time. Surveillance makes it possible to think about disease in terms of collective social behavior. Healthy individuals permit surveillance on the basis that the mechanisms of security will work against deviant cases in order to maintain the "normal" condition of the population.[32]

Biosurveillance practices also build networks that connect local, national, and global populations, offering the possibility of quantifying the health status of the world at any given moment. Surveillance animates disease by showing how outbreaks move through space and time, providing a map for intervention or prediction. Surveillance technology thus facilitates a belief that it is possible to act against disease even before it becomes visible in a population. If bioterrorist attacks can be deterred by demonstrating that an attack would have little impact on a well-prepared population, as suggested by a congressional committee in 2008, then effective deterrence hinges on the ability to swiftly detect an attack. Cutting-edge surveillance technology offers that hope.

Where CDC traditionally has distributed disease expertise to populations through local health departments, the expansion of its surveillance capabilities has increased its role in gathering data. Along with BioWatch and BioShield, the BioSense program was part of the George W. Bush administration's three-pronged attack on bioterrorism. Established in 2003, BioSense aims to collect "real-time" data from local health centers with the goal of detecting anomalies that would indicate an unusual disease event, whether a terrorist attack or an outbreak of food poisoning. Using an Internet-based platform for reporting and tracking information, the program attempts to standardize data from the myriad reporting systems currently used by local health organizations. With a price tag approaching half a billion dollars, launching BioSense was costly, and the additional labor it demanded of local health departments made it immediately unpopular with public health practitioners. Just one year after its full implementation in November

2011, BioSense underwent a massive redesign and was relaunched as BioSense 2.0, and by 2016 plans were in place to retire the program in favor of a cloud-based reporting system developed for and used by the Department of Defense.[33] CDC claims that the benefits of "real-time" information not only serve national security issues but also empower local departments to communicate more effectively with each other to better serve their populations. While local workers learn new platforms and gather and report data, CDC analysts map the patterns of normal incidence of disease around the nation and sound an alert when they spot abnormalities. Then local health departments must promptly respond to such alerts, further taxing a workforce that already complains of being short-staffed and undersupplied.[34]

Surveillance is not a passive technology. Engineers of surveillance devices make calculated decisions about what information should be collected, how it is materialized, and which people and institutions have access to it. As in science research, surveillance designers make decisions about everything from which organisms or symptoms to monitor to calibrating how sensitive the system should be to aberrations. A surveillance system also must be programmed with a norm and a way to recognize specified types and degrees of deviance, relying on both science expertise and protocols to identify an event as real. These calculations may be driven as much by the bottom line as by science. As Snider explained:

> We have to set up systems sensitive enough to pick up the organisms without spending taxpayers' money on false alarms. At the end of the day, we have to err on the side of having a civic system that's sensitive, but not overly so. There's pressure from Congress to develop those types of systems. We're developing several systems. I am not so arrogant as to say we've got it, but we're getting there. We're getting closer to not one, but a suite of systems.[35]

Snider's explanation reveals the government pressure to apply surveillance information on behalf of citizens. Surveillance technologies produce data, norms, and aberrations that may prompt swift, militarized action on the population by health and security authorities. Centralized biosecurity programs like BioSense transform local data collection into a national effort. In creating a Foucauldian population, defined by shared susceptibility to disease, surveillance identifies risk and determines the level of action expected by the governing entity. The consolidation of power at the national level builds authority over the calculation of risk, giving the state the ability to rationalize extreme interventions on the basis of the surveillance data.[36] Locating health surveillance at the national level also empowers governments to define both the scale of the risk and the appropriate response, a concentration of biopower that remains invisible to those who do not participate in the production of data.

As with other bioterrorism preparedness programs, public institutions rationalize surveillance activities by citing their multiple uses and synergistic public benefit. As the acting director of CDC's Coordinating Office for Terrorism Preparedness and Emergency Response, Dan Sosin argued that surveillance activities bound by the same rigors of science as CDC's laboratory activities could sustain the dual-use platform of bioterrorism preparedness:

> We need effective systems of gathering and sharing timely health information to lead to quality improvement and effective decision making. That requires scientific emphasis on identifying the appropriate metrics of interest for surveillance, both from the standpoint of improving the provision of care, whether that's clinical care or public health care, and for measurement in quality improvement. This focus on the science of gleaning health-related information from real-time electronic sources, and how we make better use of that, both in targeted ways and in ways of discovery, are an important place in which we can advance multiple objectives of public health at the same time . . . and that can be driven by science.

Bioterrorism preparedness goals change the types of surveillance data collected and, in turn, the surveillance technologies themselves. It creates the potential for the system's objectives to favor enabling a rapid response during an event over identifying and addressing long-term health trends. Still, Sosin describes surveillance as a "leverage point" in biosecurity, something that will affect the outcome of bioterrorist events, "if we invest appropriately and apply a rigorous scientific approach to the extrapolation of that knowledge."[37]

The expansion of CDC's surveillance programs exemplifies how existing technologies find broader application and popular support during a crisis and then are embraced along with a particular system of governing. In the post-9/11 security society, disease surveillance practices performed by CDC for decades are being expanded and repurposed. These changes have consequences for governing systems. David Lyon and Didier Bigo argued that such an expansion disperses the mechanisms of surveillance and the role of surveyor among many actors whose importance is elevated by a crisis. Moreover, because surveillance systems are perceived to be passively watching the population, they are characterized by a lack of accountability. And, as with scientific study, because the enemy is unknown, the surveillance response is gauged by available solutions rather than by the threat itself.[38] When they are under surveillance, citizens assume a new identity grounded in risk, built around a normalized population and the perception that they are living in a state of exception. Health surveillance technologies in particular characterize all members of the population as potential victims of disease.

Surveillance cannot predict when or where a bioterrorist might attack, but it can promise to be ever watchful, looking for aberrations in the normal state of the nation that might indicate disruption. A list of eight needed changes identified during a comprehensive CDC review after 9/11 included "improved" surveillance and "automated syndromic surveillance systems."[39] With new surveillance technologies, not only are humans continually watched, but our doctor's visits, Google searches, and flight plans are scrutinized for any signs of microbes. These behaviors are examined for patterns alongside the behaviors of millions of others, and the data are archived at CDC as the basis for identifying a potential attack. The surveillance process sees all life as suspect. As dramatized in the congressional report of the Graham Commission, a bioterrorism taskforce, all members of the population live every moment as near victims: "In every terrorist strike anywhere in the world, to every innocent life lost must be added thousands more who were just hours away. . . . In those moments of danger, we are all, first and foremost, citizens of a world at risk."[40]

RISK COMMUNICATION

In 2007, CDC revamped its public website, CDC.gov. It claimed the redesign was based on "science, best practices, and evidence-based research" to better serve the nine million "diverse customers" who search the site for health information.[41] Although CDC's traditional customers are state and local public health departments, the agency's expertise is increasingly sought by other government agencies, the press, and the public. CDC's online makeover followed the recognition that individual citizens were going directly to CDC to find answers to their health questions. Snider commented on this shift: "For many years we thought we were a wholesaler of health information, but now we realize that we are both a wholesaler and a retailer. It took a while to realize we were in that role, and now we are making adjustments to be a good retailer, too. It's a new concept for us: we hadn't been trained to do that."[42] Digital technology enables individuals to access CDC for health information, including information about bioterrorism preparedness, at any time and from almost anywhere in the world. Yet there are limits. Although the CDC's public website seems to offer subjects the ability to make decisions regarding their own health, it provides laypersons not with the raw data collected by scientists, but with a crafted message about how to manage their behavior. In packaging and displaying information for the public, CDC makes deliberate decisions about enticing consumers and producing citizens who will govern their own behaviors.

Many of CDC's disease-control messages have remained consistent since before the emergence of bioterrorism as a national threat: wash your hands, cover your mouth when coughing, cook your food thoroughly. However, as the agency communicates directly with citizens, it builds its sway over the health of the population. While people still seek health expertise locally and individually, the technologies of the digital age seem to offer access to a large, distant group of experts. Again, the national agency increasingly assumes a role previously occupied by local health authorities.

CDC is motivated by "the belief that the government can and should do something to improve the lives of citizens."[43] This statement by Jim Curran, now dean of the Rollins School of Public Health at Emory University and former head of CDC's AIDS program, elaborates the goal of public health to articulate risk or "redefine the unacceptable." For example, scientists can present data showing that smoking causes cancer, but the public health debate lies in what should be done about it and how much the government should attempt to frighten the population in order to bring about a social effect. Public health also places the most intimate personal behaviors in a social context where they can be governed in the interests of society. While the science of public health identifies causes of harm, and surveillance establishes the parameters of unacceptable behavior, it is risk communication that shapes public opinion in order to effect change. CDC's use of science knowledge to influence citizen behavior is a form of biopolitical governance eminently useful to the security state.

Beginning in 2002, CDC introduced risk communication into its semiannual training for Epidemiological Intelligence Service (EIS) officers, the field staff who respond to disease events. The course teaches strategies for building trust, nonverbal communication techniques, how to stage a public exhibit, how to run a public meeting, and media strategies.

The course creators define risk communication as "a science-based approach for communicating effectively in high concern situations," correlating effective communication with positive health outcomes. The possible effects of poor risk communication include "demands for unneeded treatment, disorganized group behavior (stealing/looting), bribery and fraud, . . . unreasonable trade and travel restrictions, . . . misallocation of limited response resources, . . . [and] increased disease and death."[44] As one trainer argued, "Perception creates the reality of the issues."[45] Further, the strengthening of CDC's media corps since the 2001 anthrax attacks reflects the agency's commitment to the careful crafting of media messages during times of emergency.

Although communication about the behavior of certain microbes or a preferred response intensifies during a crisis, the CDC works every day to prepare the population to receive such messages. The agency regularly publishes preparedness plans that describe specific actions people should take before and during a disease emergency. These publications advocate covering one's mouth when coughing or sneezing, isolation of a sick person, home quarantine, closures of schools, cancellation of public gatherings, and hand washing.[46] Some disease specialists argue that these types of nonpharmaceutical interventions do little to prevent transmission and even that the use of such measures "could have substantial indirect and unintended consequences."[47] On the other hand, Lisa Rotz, director of the Office of Bioterrorism Preparedness and Response, claims that such communication engages individuals in emergency preparedness: "If they feel like you're responding and you're thinking about these issues, they're going to feel a lot less anxiety. They know what's going on and how you'll respond to it."[48] Thus CDC accomplishes its preparedness mission by repackaging familiar disease-prevention behaviors as acts to secure the nation and by representing individual health decisions as community-minded actions. The institutionalization of such communications during "normal" times changes the structures through which CDC communicates all messages about disease risk.

Risk communication at CDC is also concerned with upholding its expertise. Rotz's remarks show the agency's deliberate effort to assure the citizenry that the government is at work solving the problem while also emphasizing the importance of citizens' cooperation. The government must assure its citizens that it has anticipated and contained threats to the population without interfering with the day-to-day work that sustains society. Aihwa Ong describes how neoliberal governments center on individuals by using technologies that "rely on an array of knowledge and expert systems to induce self-animation and self-government so that citizens can optimize choices, efficiency, and competitiveness."[49] Risk communication practices create just such opportunities for citizens to gather knowledge and self-animate in the face of biological threats. Together, the science, surveillance, and communications technologies used by CDC build expertise that defines the risks of bioterrorism and enable governments to create citizens who police their own behaviors in ways that may limit the movement of microbes.

OUTCOMES: STOCKPILES AND NETWORKS

Both the scientists and the CDC officials I interviewed expressed the difficulty of showing people what federal bioterrorism-preparedness funds are buying and of convincing them that the nation is prepared for biological attacks. In the absence of a bioterrorist event offering the opportunity to demonstrate an effective response, agencies may point to tangible indicators of preparedness, such as stockpiles of pharmaceuticals, or evidence that the funds are providing a social benefit beyond preparedness, such as increased numbers of health laboratories. Significantly, CDC allocates most of its bioterrorism budget to support biosecurity measures at the state and local levels, creating dual-use capacity at all tiers of public health. As the agency stockpiles vaccines and hosts private-sector training, it is becoming a national repository for goods and information, where expertise travels with pharmaceuticals, and CDC's acts have the potential to sway markets. The Strategic National Stockpile (SNS) and the Laboratory Response Network (LRN) exemplify how material changes to the nation's health system have brought public health more forcefully into the domain of national government.

The Strategic National Stockpile

Biosecurity funding has changed the expertise-only model at CDC by building the Strategic National Stockpile, a reserve of vaccines and other pharmaceutical supplies stored for biological emergencies. Using SNS resources, CDC claims it can deliver fifteen semi trucks full of medical supplies to any location in the contiguous forty-eight states within twelve hours. Working an "adrenaline junkie's dream job," SNS workers excel at logistics and have the technological and personal connections to transport items rapidly.[50] While local and state agencies are responsible for the distribution of supplies to the population at the site of an emergency,

FIGURE 15. Citizens rehearse the distribution of medical supplies from the Strategic National Stockpile. Photo: Centers for Disease Control and Prevention.

the stores are bought and managed by the federal government, giving CDC its first large-scale physical product: millions of doses of antibiotics, vaccines, and other countermeasures to biological attacks. Because this is the largest collection of such goods in the United States, CDC's decisions on what drugs to purchase, from whom, and how often, affect the pharmaceutical industry. Prior to the establishment of SNS, local health departments were solely responsible for acquiring disease treatments, even in emergencies. Now, although these groups play a vital role in emergency response, the distribution and administration of pharmaceuticals and commodities other than small local reserves are controlled at the national level. Because CDC discourages individuals from purchasing or storing vaccines and antibiotics, citizens must rely on government to provide appropriate antidotes and to distribute supplies efficiently and equitably. The SNS binds citizens to the state by their need to access care during a crisis. By controlling the countermeasures, the national agency magnifies its role during a bioterror event. The caring act of governance is expressed as the timely dispersal of drugs.

Taxpayers spend about two dollars per person per year to keep the CDC stocked.[51] Through the strategic distribution of SNS supplies, planners anticipate that they could effectively impede the spread of a bioterrorist agent. Their strategy, however, relies on local communities to distribute the product to their citizens according to an established emergency plan. The CDC rates each community's plan according to national standards, certifying the readiness of citizens to receive help from the federal government. In this new politics of biological citizenship, communities must demonstrate readiness to the national authority in order for their citizens to receive care. Thus, community-level governance operates under federal sanction in the new security state.

In addition to its expertise on biological agents and vulnerable populations, CDC is also developing expertise in managing and distributing resources. On a tour of the EOC, CDC's Captain Ralph O'Connor recounts a request from the Federal Emergency Management Agency (FEMA) for CDC assistance during a hurricane, when the cots in their medical tents were too small to accommodate overweight patients. CDC devised a plan to bring in piles of wrestling mats, and, through the logistics work of SNS, was able to locate and ship the mats to the site.[52] Working daily from a room connected to the EOC, the SNS logistics experts will be close at hand when an incident occurs, reflecting the increased use of CDC's expertise during disasters and a new base of authority for the security apparatus.

Laboratory Response Network

Though the public has no firsthand knowledge of the SNS and its location, which moves regularly as a security measure, it is easy to visualize a vast national medicine cabinet stocked with antibiotics and vaccines as a tangible product of bioterrorism funding. Rotz argues that the Laboratory Response Network (LRN) has similarly high profile because people can imagine a network of laboratories with standardized diagnostic capabilities. Established in 1999, the LRN reaches every state and consists of about 150 laboratories, including public health, military, environmental, and veterinary labs. LRN labs use a standardized platform and specially trained personnel to test for biological agents. In theory, the LRN enables a faster bioterrorism response by facilitating local diagnosis of bioterror agents. LRN directors cite other benefits of such a system, particularly in times of emergency, such as having employees who are used to working together and who use the same equipment.[53]

CDC directs the LRN by developing protocols for the laboratory response, training personnel, and determining which assays and equipment are used in local labs. Rotz points out that a network functions only if all the components and connections are maintained, a task consistently constrained by funding. CDC's work is federally funded, and LRN labs have access to additional federal funding through grants, but for the most part individual states allocate funds for the training and equipment that keep the labs on the national network. While the LRN spreads out the work of identifying biological agents, the CDC retains authority over the laboratory system at the federal level by structuring the operations and physical spaces of the network.

LRN bioterrorism funding and CDC's recognized expertise in diagnostics have created a new market for biological field equipment. Stephen Morse pulls a weighty, spiral-bound catalogue from a shelf of biology journals in his office in CDC's Division of Bioterrorism Preparedness and Response and flips through hundreds of glossy page of biotech equipment. When local emergency response teams collect their bioterrorism preparedness funds, they turn to volumes like this to purchase devices that claim to diagnose anthrax on the spot or detect foreign pathogens in the air. Morse says that much of the equipment offered here is inadequate or unnecessary, but that people with budgets to burn are buying it up because the technology is available. As a scientist and a public health worker, Morse prioritizes standardization, warning that in an emergency poor fieldwork could flood the public health system with hundreds of thousands of unusable samples. The LRN seeks to counter the forces of a booming market in bioterrorism preparedness equipment by focusing on uniformity in emergency response tools and procedures. Again, under the auspices of creating a timely response, CDC has increased its authority over local public health work, and by standardizing the equipment on the counter and the assays in the refrigerators of laboratories around the country, it is materially changing the local face of public health.

CONCLUSIONS: THE CITIZENS' CDC

"Having a plan doesn't mean you are prepared, because you haven't shown you can implement it." Susan True explained this challenge of preparedness to top-level planners from state and local government during the Meta-leadership Summits for Preparedness presented by the CDC Foundation.[54] At these summits, CDC reached out to groups who partner with scientists and government workers during crises, building, in the words of Julie Gerberding, "connectivity that includes people who are not like us—people in business, people in the faith-based sector, people in the health care delivery system, people in the nonprofit community."[55] In response to the concern that groups "not like" the CDC may have competing goals during an emergency, the CDC organized the summit to coordinate efforts by these groups before the event. A second objective was to promote the public responsibility to sustain the CDC's preparedness response.

Established by Congress in 1995, the CDC Foundation represents the continually changing face of public health in this country, showing the blurring of boundaries not only between local and national care but also between private, corporate, governmental, biological, and scientific enterprises. The CDC Foundation operates with federal funds in order to raise money for CDC and return those investments back to the agency. In fourteen years, the CDC Foundation invested $170 million in CDC health programs, drawing large and small donations from corporations and private citizens.[56] The claim that a nongovernmental organization can help CDC do better work, faster, underlies the foundation's mission, particularly with regard to emergency preparedness and response.

In fundraising, the foundation balances a booster message about the good works of the CDC with a warning about weaknesses that might be overcome through private-sector financing. Deficiencies of communication between government and private entities during health emergencies, for example, can be overcome through the leadership summits that strengthen community relations. Another oft-repeated anecdote tells of CDC epidemiologists working in the field who needed supplies not found in their kits. To circumvent the slow process of government purchasing, the CDC Foundation issues field workers with credit cards that they can use to buy rope from a hardware store or print a sign at a copy shop. The cards are funded by private donations through the foundation's website, offering an easy way for individuals to contribute to CDC's preparedness work. Though the options for participation are skewed toward monetary donations, the CDC Foundation promotes the broader idea that citizens working outside government can improve the agency's effectiveness. "After working so long in government and dealing with the bureaucracy, I had to ask, 'Can it really be this easy?'" reflects the foundation's vice president for programs, Kevin Brady.[57] Although individual citizens of the security state are discouraged from stockpiling medical countermeasures or directly contacting CDC's science experts, the CDC Foundation provides an easy way for private funding to shape the bioterrorism response.

Like the CDC itself, the CDC Foundation profits from the widespread perception of a biological threat. The ongoing, preemptive nature of bioterrorism preparedness ensures the continuing engagement of the public health system and encourages citizens anxious about their personal health to turn to government resources and expertise. In an endless war against an invisible threat, when governments cannot prove that their preparedness work has prevented an attack, the material evidence that the government is taking action offers reassurance. As Dan Sosin explains, bioterrorism

> isn't a disease that we conquer or that we tire of and then go on to another condition. In this instance, it's really the infrastructure that supports high-profile, high-impact kinds of responses to these types of events. Inevitably, we will continue to have them over time, and depending on how successfully we execute the resources we receive, and how well we measure and account for them, those resources are unlikely to dissipate in the grand scheme over time.[58]

Sosin sees no end to the funding for bioterrorism preparedness, provided the agency can demonstrate to the public that it is sustaining the infrastructure for a security response. Thus, CDC is incapable of eliminating the threat of bioterrorism, for the institution exists not to "conquer" disease but to build security infrastructure. In the face of endless threats of disease, the security state is always in search of additional technological fixes, since security mechanisms are rendered ever more vital to act against unknown and imagined threats.

Fifteen years after CDC experts flew to Washington to assess the presence of anthrax spores in a mail room, the agency seems confident in its bioterrorism preparedness mission. The director of the Office of Public Health Preparedness and Response, Stephen Redd, opened his introduction to a report titled *2015: National Snapshot of Public Health Preparedness* by describing the response to the 2001 attacks as "nothing short of heroic. Responders and citizens alike offered their best to save lives and prevent further harm. Government stepped up too, investing in critical infrastructure and seeking innovative ways to prepare the country against future emergencies."[59] Redd relates how the events of September 2001 catalyzed a response that prioritized biosecurity within the CDC. The 192-page report sustains this optimistic tone, laying out line by line what annual budgets exceeding $1 billion can buy for terrorism preparedness, including a network of local health departments funded through the Public Health Emergency Plan (PHEP) to comply with CDC's established preparedness parameters.

The document identifies one failing of CDC in the post-9/11 world: "administrative preparedness." The report explains that the agency named this category of preparedness during the H1N1 outbreak in 2009, when responders identified gaps in "the process of ensuring that fiscal and administrative authorities and practices (e.g., funding, procurement, contracting, hiring, and legal capabilities) used in public health emergency response and recovery are effectively managed throughout all levels of government. Administrative functions are the foundation of emergency response."[60] To bolster administrative preparedness, the PHEP added it as a score on its evaluation of its grantees. In May 2014, agencies staged a tabletop exercise with financial clerks and human resources teams to study how administrative work changes during a state of emergency. This new attention to administrative preparedness focuses on everyday tasks that must be rehearsed to ensure the continued functioning of government during a crisis. Bioterrorism preparedness is clearly not just the work of scientists and emergency responders but of office workers, too. The pursuit of security is the project of the whole nation, from the president to the receptionist.

The barriers to preparedness that CDC aims to break down through administrative-preparedness exercises clearly align with the project of the CDC Foundation to use private monies to circumvent bureaucratic delays during a time of crisis. The assumption that systems will collapse during a crisis, leaving field agents unable to access essential supplies, strengthens the resolve to preserve government right down to the paperwork—perhaps as a sign of normalcy and reassurance to the citizenry, or perhaps as another mechanism to demonstrate national preparedness. In a state of emergency, government capabilities must "be accelerated, modified, streamlined, and accountably managed."[61] The administrators of government, including the CDC, must therefore rehearse their response to become swifter and more efficient during the anticipated future event.

The work to build national security against biological threats can be accomplished only through the technologies that make security visible. CDC's bioterrorism preparedness practices give public face and common form to fears of germs and disease. The promise that ever-expanding infrastructure and expertise can mitigate the never-ending bioterror threat and protect citizens' lives rationalizes the continued expansion of government-sponsored programs that manage health and human life. Furthermore, waging the war on terror in the domain of public health centralizes the mechanisms of care, binding local populations to the nation-state through access to health resources. Finally, bioterrorism preparedness militarizes the daily rituals of health, reproducing individual acts like hand washing as critical acts of national safety and security. Bioterrorism-preparedness activities at CDC exemplify how closely the national security state is entwined with new understandings of biology and disease that enable governments to access citizens at the biological level and recruit their participation in securing the biological future of the population. In the new security regime, all citizens act as agents in producing biosecurity.

5

Simulation Science

Securing the Future

Every morning, residents of Playas, New Mexico, a former company town just north of the Mexican border, begin their day as many suburban families do: eating breakfast at sunrise, shuffling kids to the school bus, and then heading to work—where they are kidnapped by terrorists and held for ransom with bombs strapped to their chests until the police kick down the door and lace the air with tear gas. This scene may play out a dozen times in a day, until the day shift ends and night workers come in to patch walls and repair the broken doors, returning them to their hinges so they can be tackled again in the next day's attacks. Helicopters shuttle people and supplies to the remote location, and a guard stands in a booth out by the highway, turning away visitors and tracking the movements of bodies in and out of town. The community of Playas experiences daily the sights and sounds of war, blending blasts and explosions into the vast desert landscape and sacrificing serenity for the sake of patriotism, security, or a daily wage.

Former employees of the largest copper smelter in the region, these Playas residents elected to stay in their homes when the copper industry tanked and the company town was turned into a terrorism training center. Now, rather than working in the mining business, residents stage simulations for groups who come to the Playas Research and Training Center to rehearse their response to a terrorist attack. Through role-playing and scenario enactment, the employees simulate what terrorism might look like in an American suburb, inviting emergency planners and first responders to practice what they would do if they found an envelope of white powder at the local post office or a terrorist took hostages at the corner bank. Playas has been staged to present terrorism for modern audiences through a performance in which the audience can gauge its own preparedness to live in a world of terror.

Established in 2004 with a $5 million grant from the Department of Homeland Security, the Playas Research and Training Center marks yet another kind of place created in the service of national security. New Mexico Senator Pete Domenici lobbied for the purchase and development of this site, contending that the center would

> dramatically increase the technical capacity of emergency response organizations to manage incidents involving chemical, biological, radiological, explosive, and environmental agents. . . . The Playas purchase would add significantly to the DHS infrastructure by providing a working town for real world training scenarios . . . to handle agricultural and biological outbreaks that could significantly harm our citizens and crate [*sic*] chaos in our agricultural sector.[1]

Like the modern bioscience laboratory, the training center is secretive and isolated from surrounding communities, and it claims the status of a research facility to generate understanding of the world. The work done here similarly creates the national security state through the production of knowledge.

Current anxieties over terrorist attack recall the climate of fear experienced by many Americans during the Cold War. Through propaganda and policy, the United States produced a fear of nuclear attack that demanded a preemptive response to protect citizens from a perceived future risk.[2] Rather than fighting an enemy in combat, military strategy involved building infrastructure and implementing programs to anticipate an attack and increase civilian survival. The civil defense program also called for individuals and communities to rehearse their responses to a nuclear attack. Participating in these exercises, down to the duck-and-cover drills performed by schoolchildren, was a patriotic duty, a part of public life during the nuclear age. Joseph Masco described reenactment as a "formidable public ritual—a core act of governance, technoscientific practice, and democratic participation . . . a civic obligation to collectively imagine, and at times theatrically enact through 'civil defense,' the physical destruction of the nation-state."[3]

Not only did these scenarios imagine a postnuclear world, but they also showed just how the nation would emerge from the apocalypse and the specific practices people would have to undertake to survive. Such public rehearsal of the crisis response continues in the twenty-first century as a state-building strategy and performance of citizenship. The framing of the war on terror, however, enables civic responses that are individual, centered on the policing of one's own body—one's own microbiome—such that the collective national body is never vulnerable to attack. As during the Cold War, rehearsal and performance gives form to terrorism in the absence of a major event, changing the calculation of risk and creating new practices of both governance and citizenship. Moreover, this modern formation of civil defense is rooted in particular understandings of nature and human-nonhuman relations that have been made possible through new scientific ways of knowing life itself. This chapter explores how enveloping modern bioscience in scenario enactments provides a mechanism for displaying biological preparedness, even as it creates new biological vulnerability for an uncertain future.

Andrew Lakoff argues that in order to achieve "preparedness," the uncertain threat of terrorism must be brought into a "space of present intervention."[4] Because modern citizens have not lived through a bioterrorist event, they must learn preparedness behaviors through other means. Whereas the Cold War had a clear enemy in the Soviet Union, and the threat was assumed to be nuclear, in contemporary crises neither the enemy nor the weapon is so readily identified. National security demands new spaces and places where the threats of the modern age can materialize, and facilities like the Playas Training Center channel national security funds into the construction and maintenance of a new security infrastructure. The influx of capital fractures old systems of science and governance, reconstituting them in ways that sustain the belief that action in the present is a necessary form of deterrence.

Tracing the genealogy of simulation from civil defense programs through natural disaster planning during the 1960s and 1970s and in the present day, Stephen Collier draws attention to the ways that enactment restructures the mechanisms of knowledge production. Because statistical models are insufficient to calculate the risks of terrorist attacks or national disasters, simulations are used to generate the logics of response. The knowledge produced by such scenarios describes particular social vulnerabilities emerging from what Collier calls "the uncertain interaction of potential catastrophes with the existing elements of collective life," which forms the basis of new forms of political citizenship.[5] Thus a site like Playas attains value because of its status as a "real" town, and scenarios rehearsed in CDC's Emergency Operations

Center are valued because the interactions of participants are authentic human relations.

Scenarios draw on the conditions of the material present and then challenge them with hypothetical conflicts, creating a calculated experiment in human behavior. Live enactments enable planners to assess vulnerability by observing the interactions of individuals and groups when challenged by conflict. Significantly, we build scenarios with the same categories of difference by which we build our everyday sociality, and the "reality" of scenario enactment hinges on including presentations of race, class, and gender, as well as a range of interpersonal relations, to generate data. These representations of the future form the basis for governance in the present, as scenario enactment informs the allocation of federal antiterrorism funding.

The establishment of centers for terrorism rehearsal shows the stabilization of dedicated mechanisms for producing this type of social knowledge. Collier suggests that scenario enactment is a mechanism for governing the catastrophic elements of nature and society that Ulrich Beck, in his much-debated *Risk Society,* deemed ungovernable.[6] Enactment, unlike more traditional statistical calculations of risk, can expand to include more general societal functions. The three cases presented in this chapter show how knowledge production through simulation has taken root at all levels of government. It is a practice increasingly accepted by the community of planners and policy makers as a science for explaining and anticipating both human and microbe behavior. Not only does this practice change the calculation of risk and the types of actions taken to mitigate risk, but the heightened use of simulation has implications for how people know the world around them. Modeled and imagined futures are not certain, but current behaviors begin to give them form. Further, when models are viewed as a script of the future, they can be used to rationalize political actions that manipulate human bodies, individually and collectively.

Simulations are taking hold in the science community as tools for explaining future nature, shaping how people interact with risk. In this chapter I explore how modeling has been adapted as science practice and how it gains tenacity in the current political climate. Predictive models shape scenario enactment. Microbes lie at the core of this discussion, along with ideas of life, death, contagion, infection, and health. I look particularly at the performance rituals of nationwide full-scale simulations and a local bioterrorism preparedness exercise in Albuquerque, New Mexico. Through these rehearsals, germs shape human behavior as people work to prepare for an imagined microbial future. Finally, I return to Playas to scrutinize the founding of a commercial preparedness enterprise as a venture to redeem a community economically threatened by the collapse of a mineral extraction industry. The promotion of scenario enactment as an up-and-coming industry indicates how far terrorism preparedness can reach into communities. Though none of these simulated bioterror catastrophes have been realized, the creation of the simulations is continually doing the *real* work of preparedness to shape society.

SIMULATIONS OF LIFE ITSELF

The rapid expansion of public health agencies in the twentieth century demonstrates the strengthening of the belief that microbial behavior can be predicted and therefore managed, and that disease prevention is part of the work of government, as evidenced by the rapid expansion of public health agencies. Scientists studying germs constructed disease as a microbial attack on human life, creating the possibility that managing the interaction between humans and germs could change the outcomes of pandemic events. Indeed, microbiology created the potential for anticipating biological futures through modeling, with a particular emphasis on the biological future of the human species.

Certainly the motivations for looking closely at microbes have not deviated far from Enlightenment ideals about science and nature. The inclination to take up the microscope to conquer germs seems a likely response to Francis Bacon's call for science to "conquer nature in action," resulting in what Max Horkheimer and Theodor Adorno describe as distancing people "from nature in order to arrange it in such a way that it can be mastered."[7] Studying microbes showed them to be increasingly not like us and therefore objects that could be dominated. William Leiss, however, offers the interpretation that the Enlightenment ideal of mastering nature was to bring about a mastery of self and control of the human impulses.[8] The impulses of civil defense align with this understanding of the Enlightenment project: rehearsing responses to crisis aims to eliminate panic and promote social order and control. The scientific inquiry of microbiologists arranges nature for the simulations that become the core public work of preparedness, binding the study of nature to the ongoing project of self-mastery.

The biologist Lynn Margulis eloquently described the work of her profession: "We observe growth and reproduction; we spy on protocist sexuality and physical maturation; we measure the responses of bacteria and protocists to environmental 'insults.' We are especially concerned with these microbes' behavior, rich social lives, and interaction with sediments as they form persistent community structures."[9] The work of microbiologists is not just to observe but to measure, poke, and prod in order to observe microbes relating to each other and their environment, to see the flows of chemistry and energy through which life continually perpetuates itself. Seeing the rich social life of a virus is to watch it become alive, because the virus does nothing without a host, not even metabolize: interaction animates viruses, whose communities (like ours) are always multispecies. Knowing a microbe therefore entails not just knowing its physical traits but also knowing how it works toward its own survival through interactions with other species, including humans.

Public health science frames this knowledge differently, asking how the interaction between humans and microbes influences the survival of the human species. This question leads to a particular type of microbial science that maps microbes onto the human population. Part of the way people understand germs, for example, is in terms of mortality rates: the percentage of people who die after contracting certain infectious diseases. While some mortality tests can be created in the laboratory, the best understanding of mortality comes from observing how a disease moves through the human landscape. The knowledge that smallpox has a 30 percent mortality rate was generated by studies of human death. By contrast, an understanding of how the virus enters into and infects cells can be acquired only by looking through a microscope. Smallpox is more broadly understood by its rates of death and infection than by the unseen behavior of the virus. This way of knowing an organism, by statistics about how many humans it could kill, is uniquely applied to germs. While lions and bears and pit bulls also end human lives every year, the interaction between a human and mammalian predator is not defined by the odds that it will kill the human, nor is the animal held responsible for a percentage of human deaths that rises to the top of every Wikipedia entry or information pamphlet about it.

The calculation of a mortality rate leads to the making of predictions. The laying of odds looks forward, anticipating how the exposure to a germ in the present will shape the future, and how many futures are available to an infected body. Observing past infections creates the ability to calculate a rate by which the microbes will travel through the population and affect human life. Assuming mortality rates remain consistent, the future relationship between humans and microbes can be described in terms of how that particular encounter has expressed itself in the human population in the past. The sample size for the calculation of mortality rates of the most deadly germs is small, however, and a range of environmental factors affect the spread and severity of disease. Populations carry differing levels of immunity, for example, and some diseases fade with the changing seasons. Human behaviors may intervene in transmissibility or increase the likelihood of survival, the objective of our "war" against germs.

The public health work of microbiology establishes the battle plans and strategic defenses in the war against germs that help the population fight against the statistical inevitability presented by the mortality rate. Perhaps because the actual work of microbes, viruses in particular, remains mysterious despite technologies that enable us to watch closely, we are more comfortable understanding microbial volition in terms of microbes' impact on human life. Perhaps because the possibility that a cold, indifferent nature could end human life is unbearable to contemplate, the human cultural imaginary has created microbes as monsters, beings with intent to cause harm and possessing all the deviousness, coldness, and stealth of a malevolent villain.[10] If we understand germs as the enemy, then the war plan must outmaneuver and outwit our microscopic rivals in order to reduce mortality and create a future that favors human survival. In war, the outcome is not certain, and strategizing is constant. Modeling disease outbreaks anticipates the action of the microscopic enemy, while simulation tests the effectiveness of public health strikes in a war against disease.

Mathematical models of disease stand alone in public health practice, but they also inform larger simulation exercises. They condense time and space and depict disease in a way that can be presented on a screen, in a chart, or through a series of graphics, making epidemics comprehensible to audiences at a glance. I encountered a model in a museum exhibit, for example, that allowed individuals to select a germ, a location for an outbreak, and a few parameters governing how people would move. The computer provided rates of transmissibility, infectivity, and mortality for the selected virus. I pressed the Go key and watched the computer paint a picture of disease cases over a world map on the screen, swiping lines across the globe when an airplane carried an infected body and changing dots from blue to orange to red as death tolls rose. As the person who selected the disease, the season, and the rate of international travel to a global event, I was a creator of this canvas. In subsequent attempts, I tried to select the parameters that would do the most harm and paint the most colorful (and grim) future world and later to discover what situations contained the disease and left the canvas minimally blemished. After a few minutes of play with the model, I developed knowledge of disease behavior that allowed me to manipulate outcomes and change how disease manifested in this virtual world. The model defined disease in particular ways and presented social effects through a narrowly defined window.

My interaction with the model taught me that disease outcomes could be dramatically changed through the management of human travel, but it told me little about seasonality or residual immunity. As a user of this model, I became convinced that the best way to contain an outbreak was to restrict air travel, because that was the primary input available to me to manipulate. Models must take seriously the limitations of the knowledge they produce, recognizing that even the most complex models represent a finite number of inputs and present deterministic outcomes.

Disease models have taken a central role in preparedness planning. Models allow planners to imagine how human behaviors and deliberate interventions might change the course of a disease outbreak. While the scientists who create the models recognize them as a "useful tool to assist in assessing different policy options,"[11] most are quick to point out that they are "an aid to understanding, rather than being an end in themselves."[12] Models represent the world in a simplified way. This simplification is useful precisely because the world is complex. Models also allow manipulation that would be unethical or impossible in reality, creating "the ability to experiment with a complex system without actually tampering with the real system."[13] Conversely, because such models cannot be tested, their reliability is open to question. The alignment of a model with an experiment may initially seem unscientific, but Hugh Gusterson cautions against treating these two approaches as oppositional. Not only does scientific experimentation inform the structure of a model, but all experiments contain within them an element of modeling, through which external factors are controlled and the phenomena isolated through experimentation become the paradigm for broader explanations of the system.[14]

Gusterson also warns about the distance that modeling creates between scientists and their subjects.[15] Models of weapons simulate explosions and calculate deaths over and over again on the screen. This process isolates the weapon designer from the lethal effects of the weapon. As I fed data to the modeling program in the museum, I spent little time dwelling on numbers of lives lost in the disease pandemic

I created, instead reading the outcome as a personal challenge to create a better weapon or save more lives.

The value of models has been explained in many ways, including the capacity to verify existing knowledge of a system, to evaluate the relative importance of elements within a system, and to explain complex aspects of system behavior to general audiences. Scientists readily point to the limitations of modeling and caution that, particularly when using models to communicate about a system, the limitations of the model must be acknowledged and explained, lest the model appear "deceptively 'correct.'"[16] If, as one practitioner argued, "mathematical modeling is no more and no less than a tool to support clear thinking," then the value of a model is not in predicting outcomes but in providing tools to compare the results of the application of multiple strategies in order to understand how external factors shape systems.[17] Indeed, models contain within them evidence of the unpredictability of the world because they display many possible futures.

THE SMALLPOX MODELING WORKING GROUP

Perhaps because models address uncertain futures with present action, the political move toward bioterrorism preparedness in the early twenty-first century created an easy alignment between simulation and the core of public health practice. The security regime needed tools that left the future open to all imaginable ills but also empowered people to police their behavior as they looked toward those futures. As I watched disease spread across the globe in a simulation where I was a decision maker, I experienced this power of a simulation to connect individual acts to global futures. The developers of terrorism preparedness exercises sought models that identified key components of the simulated sociobiological system, such as vaccine distribution or quarantine, as points where decision makers could act. In the very early years of this century, preceding and on the heels of 9/11, the number of scientific articles about public health modeling, and smallpox modeling in particular, surged.[18] In response, the Advisory Council on Public Health Preparedness of the secretary of health and human services gathered a group of scientists to consider how models might be used to inform policy making and influence bioterrorism preparedness. The Smallpox Modeling Working Group concluded that modeling could be of value as a method to compare intervention approaches, to identify the assumptions and logic that gird public policymaking, and to pinpoint gaps in knowledge. Thus, they set out to create a standard by which modeling science could be made more useful to policy makers.

The committee set out to establish standards for smallpox modeling. They gathered a group of modelers, epidemiologists, and policy experts and challenged them with key questions. For example, smallpox modelers had to make decisions about the rates of infectivity they would build into their models, drawing on their own interpretation of statements like "Patients are most infectious from the onset of the anathema through the first seven to ten days."[19] Different methods of entering information like this could result in significant differences in the modeling results. Other data points included the effectiveness of vaccines, the likelihood of human behaviors, the reliability of diagnosis, and so on. Similarly, the committee dealt with definitions of public health practices, debating meanings of *isolation, quarantine, surveillance,* and *containment*, for example. Isolation from all bodies plays out differently from isolation from nonfamily members, and staying home from work differs from having no human contact.

Moreover, models simulate interactions between individuals to study contagion, and designers make varying assumptions about how bodies mingle with each other in the cultural landscape. Models that presume limited mixing, or mixing within similar social groups, create a different epidemiological future than those that model diverse interactions between wide-ranging parts of society. In their model of smallpox in two rural counties, Joshua Epstein and colleagues eschew "all homogenous mixing assumptions at any level."[20] They establish parameters for human behaviors that are neither random nor homogeneous but are governed by rules and hierarchies selected by the scientists, such as singling out hospitals and families as social units that play prominent roles in smallpox transmission. Such assumptions translate a scientist's judgments about how people mingle between social groups, developed through scientific evidence or personal experience, into data describing how microbes move within a population. Thus a model assuming insular communities might show a rapid spread of disease within the community and limited transmission beyond, while the opposite might be true if the parameters of the model are set to show high social mixing.

Models may thus inscribe differences of race or class in the landscape, which affect how political actors and citizens themselves view disease risks. Throughout human history, assumptions about the prevalence of disease in poor and minority communities have been defined more by social beliefs than by biology. Immigrants to the United States have been quarantined for plague, typhoid, tuberculosis, and smallpox, exemplifying how public health systems advance the systems that oppress outsider groups. When the tools of science are brought to bear on the reproduction of difference, we see troubling social effects, such as the eugenics movement or the HIV/AIDS crisis.[21]

To combat bias and simulate reality, scientists try to generate uncertainty through the careful scattering of random human behaviors. This is as close as they can come to turning avatars into agents who act individually. Including parameters for social behaviors in the model is essential because governments use models to find ways to change human responses. The decision to randomize human behavior undermines the rationale of using the model to influence human behavior in the first place, but it reminds us that modeling exercises for public health scrutinize social interventions and human interactions as material expressions of the movement of microbes.

The microbes, too, are subject to this coding. Ellis McKenzie's report on the activities of the working group relates the particular challenge they faced in agreeing on the "facts and hypotheses about mechanisms driving the dynamics at almost every level."[22] Debates over smallpox biology revealed as many gray areas as debates over human social behavior did. For example, each subtype of smallpox has different rates of manifestation, death, and transmission, and the group had to agree on how each subtype would be used in the model. The data were selected by committee, which, after vigorous debate, came up with a set of numbers and equations that prescribed how smallpox would function in the system. Coding microbial behavior creates the illusion of scientific certainty. The modeling mechanism demands that nature be knowable, and when a group of experts comes to consensus about that knowability, the consensus sustains a belief that microbial behavior can be predicted.[23]

The models depended on historical smallpox data that were gathered primarily for the purpose of eradicating the disease, not to guide models of the smallpox-human interaction in a bioterrorist attack. Because the disease has been eradicated, no new empirical data can be created about its spread and infectivity in a population. Data available for other diseases concerning transmission rates, vaccine efficacy, and fatality rates, and how the microbe might engage with the current U.S. population, are not available for smallpox: they can only be hypothesized. Modeling is a mechanism for producing new authority, but the smallpox models imagine futures based on increasingly distant pasts and shaped by cultural ideas established during a centuries-long war with the disease. The smallpox virus will be created again and again through mathematical extrapolations of past behaviors, building certainty through repetition and trust in scientific expertise.

The creators and users of these models recognize that they are incomplete and imperfect shadows of the real but still see them as means by which we can better perceive and understand reality.[24] While scientists may use models to explore complex questions and study the present, policy makers use them to guide decisions about taking action. Models may not perfectly mimic reality, but they still shape it. The final objective of the Smallpox Modeling Working Group was to use a system of testing and peer review to evaluate the effectiveness of the modeling tools for use in guiding policy. The scientists concluded that standardizing input parameters and using multiple models would increase validity and build consensus around certain depictions of the system. Though the outcomes of each model were different, the team agreed that surveillance and containment consistently changed the outcome of the smallpox attack: they therefore advised policy makers that these measures should be used to counter the effects of a smallpox weapon. They also recommended that policy makers continue to use modeling to plan for bioterrorist events, so long as they acknowledged the limits of a single model and controlled parameters to create consistency between models.[25]

UNCERTAINTY

Models have been offered as one tool for overcoming the relative lack of experiential knowledge of bioterrorism attacks. The institutionalization of modeling practices within the working group and through other public health modeling has the potential to reshape how we know disease. Models demand standardization and certainty about both microbial and human behavior, but their clean and tidy conclusions mask the complexity and uncertainty of the decision-making process that sets those parameters in the first place. For this reason, as Sherry Turkle observes, "simulation makes itself easy to love and difficult to doubt."[26]

Models matter because they create knowledge that delineates the futures open to biological citizens. Gusterson suggests that examining how "scenarios of the future are modeled, evidenced up, and presented to the public" reveals a particular class of anticipatory knowledge.[27] The practices of modeling the future and then communicating that future to public audiences are often scientifically ambiguous and politically influenced. The public alarm over bioterrorism in the late twentieth and early twenty-first centuries along with the boost of federal funding, motivated the expansion of disease modeling, and situated this work in an intensified political context that shaped how people designed studies and applied outcomes. At the same time, the development of modeling as a scientific process created an ideal tool for the security state, which uses anticipatory knowledge to build expansive futures.

In their argument for an individual-based model for smallpox bioterror, Epstein and colleagues claim that without models, there is no way to "gauge uncertainty."[28] Despite the streamlined depiction of a modern biological society, the ability of a modeler—or even an end user like me in the museum—to manipulate countless factors in the system and create seemingly limitless outcomes produces the future as endlessly uncertain. This uncertainty is the currency of the modern security state. Further, reducing human biosocieties to agent-based systems represents the future as the inevitable outcome of whatever interventions are taken in the present. If populations can be assured of avoiding outcomes X, Y, and Z by taking actions A and B, the model might convince people of the value of the project of governance to grapple with uncertainty and ensure certain outcomes.

The implications of increased public health modeling extend beyond governance. Models change how knowledge is made. Modern understandings of dynamic, evolving nature—created in part by microbiologists who built a vast, agential, inexplicable world right on the surface of the tangible world in which we live—challenge the Enlightenment ideal that we can gain complete knowledge of nature and ourselves through scientific study. Modeling emerges from the realization that scientific study of the past still leaves the future unknown. Despite decades of experimentation and observation, scientists can't fully explain how viruses live, and the presence of viruses like HIV, West Nile, or influenza create constant uncertainty in the population. The model aspires to make uncertainty knowable. When scientists adopt new methods for explaining the world, science is transformed.

The modeler's quest to mimic human behavior by applying both randomness and order mirrors broader scientific thinking about nature after Darwin as a hierarchy of external factors that act on a subject. By working to account for human agency, disease models may naturalize human behavior. An individual's decision to shop at the mall is input as one parameter among many, along with data about microbial infectivity and contagion. The biological subject, perhaps symbolized by a dot that changes color with infection, is transformed through an inevitable and predetermined progression of disease: the body "naturally" moves through stages of infection even as the subject's behavioral responses proceed according to established social and biological parameters. Members of the population in a disease model are categorized as susceptible to infection, infected, infectious, or removed. This label determines the outcome of their social interactions and how their bodies shape the world around them. Because the scientific method calculates variance in the biological world in advance, the biological parameters of a model have predetermined ranges of variability. Experts evaluate a model's level of "biological realism."[29] If the model offers an acceptably accurate representation of microbial activity and the external factors are sufficiently random, the end user can press the Go button and watch the world be remade by a single biological shift created by a germ.

The Smallpox Modeling Working Group was created with the explicit purpose of determining how modeling works within a suite of preparedness practices. After analysis, the group recommended the continued use of models to plan bioterrorism exercises, deeming them to be an effective mechanism for identifying human interventions that might affect outcomes and could be rehearsed.[30] Modeling builds a convincing argument for the management of human bodies to mitigate the spread of disease, for a grim scenario can be remedied by firmer control over human behaviors. What changes with a few clicks in the virtual world, however, may be impossible to bring about in a dynamic sociocultural landscape. The solutions offered by the model, along with the core mechanisms through which modeling operates, create expectations that human behavior must change.

BIOTERRORISM ROLE-PLAYING

In 2008, the United States paid $12 million for a terrorist attack. The Department of Homeland Security awarded a contract to Northrop Grumman Corporation to plan a simulation of a terrorist event that would affect six Southeastern states, four countries, and the Navajo Nation. As part of the National Level Exercise (NLE) program, emergency planners in these regions were assigned to try to stop the attack and keep it from crossing their borders. The NLE (formerly called TOPOFF and recently renamed the Capstone Exercise) is part of the National Exercise Program, the primary mechanism for calculating preparedness and testing the capabilities of the U.S. homeland-security apparatus.[31] The "Top Officials" (TOPOFF) series of terrorism preparedness exercises began in 2000 with a simulated biological attack, seventeen months before the anthrax attacks of 2001. Mandated by Congress in the late 1990s, the scenarios were staged every other year. They consisted of large-scale terrorism preparedness drills involving decision makers at all levels of government and from the private sector. The scenarios lasted about a week, took place in multiple sites, and involved citizen participants playing the roles of victims, attended to by local emergency responders. More than fifteen thousand people participated in TOPOFF 4 in 2007. In the absence of a major terrorist attack, these scenarios enable people to imagine what might happen if weapons of mass destruction were turned on their own bodies and communities.

The large-scale exercises staged in the years following 9/11 and the anthrax attacks performed a range of possible outcomes of terrorism on the American people on a national stage. The photographer Nina Berman recalls that during the 2003 TOPOFF 3 exercise, the planners prepped thousands of participants with stage makeup and props to mark the presence of disease or injury: "They had all the elements of theater there—pillows to make you 'pregnant,' fake blood, biscuits you chew and spit out like vomit—and they were pretty determined in their objective to make it seem real."[32] The scene in Connecticut began with a fireball, with an explosion loud enough that observers were issued earplugs. One observer watched the actors step into their roles: "As the mushroom cloud of smoke drifted away . . . hundreds of gory victims processed into the site to assume positions of death and agony."[33] The actor information sheet instructed participants, "If you are assigned the role of a psychologically distressed person, please act upset, not out-of-control." Participants were instructed to panic or to assist or even to die according to a script of the future authored by companies engaged by the government.

These exercises staged a biological transformation in the population. Stage makeup turned healthy bodies into bodies under attack. The participants performed victimhood and insecurity, though their security was not actually in danger—and they really seemed to enjoy themselves. They delighted in the opportunity to perform, to act out their own demise, to be injured without experiencing harm. The simulation was not a separation from the real world but a pleasurable immersion in it. The stage makeup cast citizens in roles designated by the organizers. It permitted "normal" people to behave as victims because they were actors; they did not have to act according to their prescribed social role, because the crisis had broken down the social order. The later TOPOFF

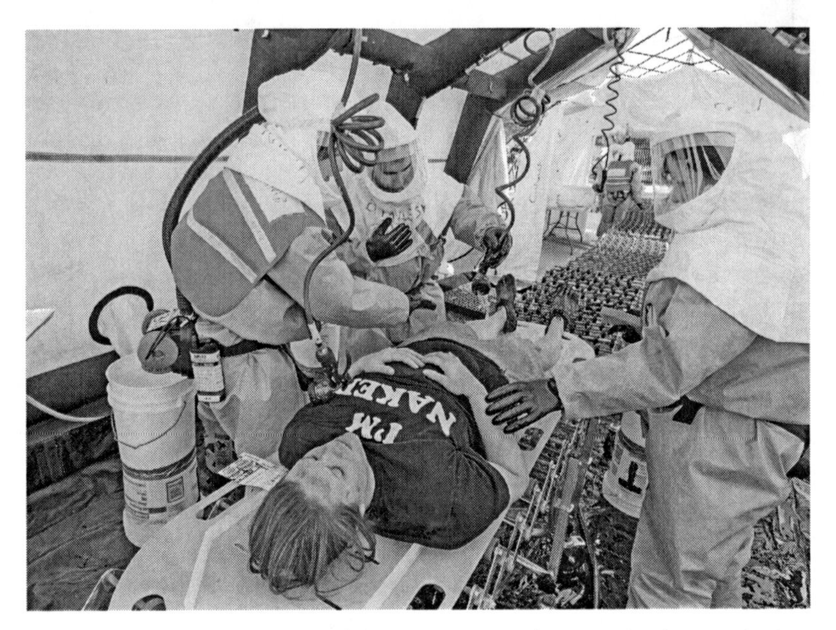

FIGURE 16. Trainees at the Center for Domestic Preparedness practice decontaminating an actor identified as a victim by a shirt proclaming "I'm naked." Photo: Federal Emergency Management Agency/Benjamin Crossley.

scenarios also featured a "Virtual News Network" (VNN) to report on the events in real time. Though the VNN component was included to scrutinize how information about the event and preparedness response could be communicated to the public, it also provided an audience for the performance (participants could play to the media observers) and deemed the incident worthy of media attention, one of the key ways people evaluate the significance of events.

Because Americans have not lived through a large-scale bioterrorist event, ideas of what an attack would look like must come through other means. Disease events have been playing out in fiction for centuries, and the Hollywood blockbuster has provided biological horror scenarios involving everything from mad scientists and government ruin to zombies and alien invaders.[34] The NLEs, however, perform terrorism locally, using infrastructure in people's communities.

Scenario enactments draw authority from their immediacy. They receive local media attention, and nonparticipants in the community are warned in advance to avoid a "War of the Worlds" panic effect, responding to the simulated events as if they were real. They hold sway in the political realm because they produce a new way of knowing human behavior. Policy makers imagine the enactment as an experiment designed to test human behaviors during crisis. In particular, the scenario is valued for its access to authentic emotion, which is missing in computerized models. Real people, even when acting from a script, are presumed to bring unpredictable emotions to the scene. Participants report experiencing frustration, fear, panic, confusion, and even hopelessness during events, despite knowing that no attack took place and the stakes were low.[35]

The materiality of these full-scale scenarios is vital to the way they create knowledge. Scenario enactment creates a materiality for bioterrorism events that have not happened, convincing citizens that the environment in which they live could sustain a biological attack. Because germs are not visible in the environment, they must be made visible in people's bodies and behaviors: scenarios imagine how a bioterrorist attack would manifest in a community in a way that demands emergency intervention. Scenarios also make germs intelligible by enacting how the movement of microbes through a population changes social relations. They allow people to imagine a moment when the slow movement of disease causes the rapid transformation of society.

Though preparedness exercises aim to generate knowledge of human behavior, they also remake microbes. A model or a tabletop exercise does not deal with real human bodies, but the enactment emphasizes the human body as it responds to infection. The assumption is that managing infected bodies in the field during a crisis is a singular experience that can be understood only through the material interaction.

Scenario enactment plays a particular role in making bioterrorism certain. As Turkle observes, "Simulation demands immersion and immersion makes it hard to doubt simulation."[36] When a scenario shows people panicking and emergency systems failing, it is difficult to doubt the authenticity of that response—even if it is scripted. Immersion generates questions that might otherwise remain hidden. Natasha Myers argues that models show science in action, embracing the uncertainty and intuition that characterizes science practice. Models are meant to be engaged with, to become technical objects that can be reworked to create new kinds of questions, to fully engross the body and mind in the pursuit of knowledge.[37] Scenarios are a form of model that engage hundreds or thousands of actors in a collective imagining of the future.

Especially in bioterrorism enactments, the unpredictable qualities of microbes combine with erratic human emotions to bring unknown futures under scientific scrutiny. Just as human actors are recruited and assigned a role, planners select microbial actors and script how infection will perform in the population. As with computer models, simulations of microbe behavior use information about the disease's infection, contagion, morbidity, and mortality to plan how many infected people will need to go to the hospital, how many people they will infect along the way, what percentage of those infected will die, and so on. CDC's recommendations for preparedness exercises urge scenario developers to include "as much complexity and unpredictability as possible, and bas[e] scenarios on what is likely to happen according to the microbiological, immunological, epidemiological, and disaster literature, not on myths or on widely embraced assumptions."[38] Note the paradoxical expectation that science can inform "what is likely to happen" even in the face of unpredictable behavior. Known unpredictability can be managed. Scenario enactment creates new systems for knowing the world by generating and studying future worlds modeled on the past. The knowledge produced through the enactment naturalizes political actions that manage human bodies, individually and collectively, in the pursuit of biological security.

ACTING OUT

The current anxiety over terrorist attacks recalls the climate of fear experienced by many Americans during the Cold War. Through propaganda and policy, the United States produced a fear of nuclear attack that compelled citizens to take action to protect themselves from a perceived future risk.[39] The civil defense programs of the 1950s and 1960s called for communities to rehearse their evacuation plans, and some drills were followed by public discussion of who would have died had the attack been "real." This pattern of rehearsal and reflection resurfaced in community-level bioterrorism preparedness exercises. A 2005 recruitment poster for the national TOPOFF 3 exercise lauded community volunteers who signed up to role-play victims of biological attack as "the most important people participating in our exercise." Enrollment materials promised "an interesting and enjoyable evening and as a result, our community will be better prepared to face real challenges in the future." As during the Cold War, discussions about preparedness also served as a space for broader debate over the powers of science over human lives.[40]

Not only is enactment a form of knowledge production that informs future policy making, but it is a performance that demonstrates to citizens that government is prepared, or at least working to prepare, while also ascribing to citizens their own responsibility for the security of the nation. In order to "emotionally manage" the population, Cold War civil defense required that citizens contemplate their own demise and the end of the nation-state, with the goal of creating citizens who would not panic in terror but would be motivated by fear to work for their own survival.[41] Preparedness practices do not aim to prevent disaster but assume that it will happen.

Because a biological event unfolds gradually, as bodies become infected or protected against disease through medical treatment, bioterrorism preparedness plans largely focus on medical infrastructure and access to bodies in order to provide medical care. Through scenario enactment, emergency responders seek to identify barriers that inhibit the circulation of needed supplies and weaknesses in the health infrastructure that would become critical if stressed by the rapid onset of disease in the population. Scenario enactment begins to shape these social spaces when we no longer build hospitals for the "normal" state of the population but instead design spaces that will function during times of crisis. Lakoff and Collier theorize that vulnerability is located within the "vital systems" that sustain collective life, such as water, energy, and transportation. They argue that Cold War civil-defense practices first established a relationship between infrastructure and event by imagining the disruption of vital systems through the unlikely but potentially catastrophic event of nuclear war. "Critical infrastructure protection" assesses vulnerabilities and then works to eliminate them. As a tool of this preparedness apparatus, scenario enactment makes the underlying infrastructure visible by simulating its disruption.[42]

Biological and chemical events are unique in the realm of war and terror because biological weapons do not collapse buildings or derail subways. Instead, the infrastructure at risk is made up of human bodies and relationships. Everyday activities that define society, the movements and meetings of individuals, create an infrastructure that is vulnerable to disruption and therefore merits critical infrastructure protection. Thus, social behaviors must be inventoried, assessed, and then modified to reduce vulnerability. This process extends the reach of government into the management of individual behavior, creating a particular authority over citizens' bodies that allows the community to be militarized as never before. Because the initial attack with a biological weapon may not be perceptible, social behavior may not initially change. This possibility heightens the sense that we may be under attack at any moment (or already) and also requires that the infrastructure be managed even when it is not symptomatic. The potential use of infectious agents further extends the timeline of the response: every handshake or sneeze on the subway may mark the onset of another event. Because human social acts transmit disease, preparedness interventions rationalized through bioterror simulation have a particular power to advocate changes in social behavior.

To justify policy changes, scenario enactments must create a level of expert knowledge about human actions, particularly during times of crisis. This expertise concerns not only the response of everyday citizens to an attack but also the work of people with authority to maintain order during a crisis. Emergency responders and employees of government agencies act during an event to manage people; from the perspective of governance, these agents must perform their assigned roles efficiently in order to secure vital systems. Their ability do so hinges on defining the relationships that constitute government in a way that enables an effective response to the crisis. Thus one of the primary objectives of enactment exercises is to draw attention to who, in all the many governing bodies that respond to a crisis, is doing what, when, in collaboration with whom, and using what resources.

In scenario enactment, coordination between emergency responders can be delineated, rehearsed, and evaluated. CDC advocates for the use of preparedness exercises because they foster "relationship-building" and provide a way to evaluate "performance" of emergency responders.[43] The Department of Homeland Security advocates rehearsal as a time to practice communication between the many government agencies charged with responding to terrorist events, characterizing such a crisis as an event that cannot be managed by a single agency.[44] The Federal Emergency Management Authority (FEMA) promotes the "collaborative, whole community approach" of the national exercise program, which involves "individuals, families, communities, the private and nonprofit sectors, faith-based organizations, and all levels of government."[45] The official reports of many exercises, whether at the national level or internal to agencies, repeatedly point to the breakdown of relationships between governing bodies. For example, the "After Action Quick Look Report" on TOPOFF 4 criticized "difficulty conducting and coordinating multiple missions at the incident sites," problems with "command structures that did not follow the National Incident Management System," and "inexperienced personnel," recommending the following "corrective" actions: "Further analysis is required to understand the reasoning behind the various Unified Commands established during the exercise and identify ways to address coordination issues" and "review and clarify SOPs [standard operating procedures] dealing with the integration of specialized response, assessment, and law enforcement teams."[46]

These reports are viewed as documents that will help governments prepare for events by pointing to areas where governance failed to meet expectations laid out by the scenario planners or established by the agencies. They also provide a basis for planning relationships and creating systems for emergency response. Finally, they motivate further training, such as the management program run by the CDC, to bolster and sustain relationships that prominently demonstrate the care of government during an event. These sustaining acts may be as simple as updating phone lists of people to contact during an emergency or as complex as restructuring jobs and roles within an agency. Scenario enactment demonstrates the importance of building relationships as an act of preparedness.

Enactments further establish the systems of contact and exchange between governments and emergency responders as vulnerable interactions. Beyond delineating authority to reduce its vulnerability, the scenario works to promote a particular type of human behavior among emergency responders. For example, one study of earthquake drills and community preparedness programs argues that scenarios are "an excellent method of teaching rapid response-style thinking, decision-making and the development of managerial skills."[47] Whether these skills could be taught in other circumstances seems secondary to the overarching idea that scenarios should teach a particular reactive way of thinking and making decisions.

The failure of governance is frequently measured in terms of rapidity of response. Time and again these scenarios end in defeat, whether a tally of lives lost or "reports" of hospitals unable to handle their real patients, let alone the infected actors. The abrupt end of the scenario allows no time for recovery, no time to imagine how the population might look in another month or year. The enactment creates an arbitrary, fictional endpoint and in so doing creates a moment to assess and evaluate performance to date, provoking questions regarding the rapidity of response. A report on the first TOPOFF exercise noted: "The state was very late to take control of medication and disbursement."[48] The state's response can be measured as timely only according to prior expectations, which are simultaneously being created through the exercise itself. If one person dies, the response can be seen as untimely, insufficient, or "late." Emergency events create a persistent present state against which all interventions react, inevitably after the event and therefore inevitably too late. The motivation to respond more quickly places the burden on responders to change their behavior in a way that allows them to anticipate the event. To reach that standard, planners must build their anticipatory knowledge and then exercise skills in rapid response. Thus, the people who participate in the scenario as part of their jobs perform a logic of governance that values quick, decisive thinking as a way to provide care.

Scenario enactments not only serve as rehearsals for a bioterrorism event but also teach participants what their roles should be. Each scenario shapes how incidents—real or staged—will be known and managed.[49] That knowledge consists of new forms of expertise grounded in "imaginative enactment" rather than experimentation or statistical calculation.[50] Expertise built from imagining the future has the potential to vary widely from expertise created by analyses of the past, particularly in bringing threats to the attention of citizens. Enactments can build expertise in responses to anything from a smallpox bomb to a zombie attack, whereas planning based on the past is restricted to mundane examples of influenza or a few cases of anthrax. Furthermore, scenario enactment shapes government practice, training actors for certain management qualities, rationalizing policy decisions, and allocating funding to fill preparedness gaps. As seen in the early twenty-first century, which has been characterized by a resurgence of Cold War–style scenario enactment, the desire to know the future in order to more rapidly respond to an ever-growing range of threats enables an expansion of government authority. Scenario enactment thereby becomes a mechanism for creating the unthinkable, along with the government to manage it.

HOMETOWN PREPAREDNESS

All scenario enactments are not created equal. Staging a full-scale exercise can cost millions of dollars. To use funding efficiently, they tend to involve worst-case scenarios, often attempting to address multiple or overlapping threats simultaneously or adding one twist after another as the scenario unfolds. Marieke De Goede theorizes that premediation, or the mapping of futures by the culture industries, deliberately operates at the outside of the realm of possibility in order to imagine the control and management of the future. "This does not mean that disastrous imagined futures will inevitably play out, but it does mean that the imagination of some scenarios over others, the visualization of some futures and not others, entails profoundly political work that enables and constrains political decision making in the present."[51] The calculation and motivation that inform scenario enactment have political roots and political outcomes. Decisions about which disasters to stage and which threats to enact have consequences in people's daily lives and further work to shape the preparedness infrastructure. In a radio interview in March 2011, Craig Fugate, a FEMA official, confounded the boundaries of imagination and calculation, explaining that the agency is building an infrastructure to target a broad, theorized event first:

> It's not so much a vivid imagination. As we look at a lot of our historical events, and we put it in the context of what would it look like today? And then we take some of the theoretical things, such as what happens if a terrorist attack occurs? What we learn is you cannot plan to what you're capable of doing. You have to plan against the events that could happen and build systems that start with that and can scale down; versus that you are going to scale up from small disasters and be successful in a catastrophic event.[52]

Fugate's remarks exemplify how political calculations grounded in scenario enactment produce an expansive response, concerned with building a preparedness system that can respond to an ever-growing range of threats. He argues that emergency planning must begin with the imagined event rather than the existing systems, with the large event rather than the small. Envisioning the apocalyptic future holds the key to managing small-scale events. This approach to emergency planning rationalizes both the practices that calculate "the events that could happen" and a broad emergency-response plan that assumes large-scale disasters before small ones.

Bioterrorism preparedness simulations are notable for their diversity in scale and focus, ranging from international to local in scope and from "full-scale" to "tabletop" in magnitude. Tabletop exercises gather a group of decision makers around a table, assign them management roles, and ask them to talk through their responses to the event. Often elements of realism, such as a news feed played "live" during the event, accompany a tabletop exercise to lend authenticity to a scripted event that collapses time and reduces the experience of bodily harm and mortality to words on a page. Dark Winter, held in 2001, and Atlantic Storm, in 2005, are famous national examples, involving former politicians playing key political actors, such as former secretary of state Madeline Albright in the role of the president. Dark Winter imagined a smallpox attack, using scientific data and parameters from a naturally occurring smallpox outbreak in Yugoslavia. The outcomes of this first major national exercise predicted far-reaching effects: decisions made at the table were input into the computer model to generate the next phase of the event. When the Dark Winter participants called it a day, the computer showed their decisions had led to more than sixteen thousand cases of smallpox in twenty-five states. Though the exercise was staged by nongovernmental entities, the outcomes of Dark Winter influenced several policy decisions, particularly following the terrorist events in the United States later that year.[53]

Increased funding for bioterrorism preparedness activities at the national level moves into communities through federal preparedness initiatives. CDC's Cities Readiness Initiative (CRI) distributes bioterrorism preparedness funding to more than seventy urban areas so they can develop and rehearse their own emergency response plans under CDC's direction. Specifically, cities must plan how to distribute the Strategic National Stockpile (SNS) medical countermeasures to their populations within two days of a bioterrorist attack. Rather than forcing a standardized plan on individual communities, the CRI approach channels money to cities to do the planning. Each city's plan must be designed to work within the local infrastructure and address the particular needs of the community. Thus, the planning is in one way very local, even as it works to connect city residents with a national supply of pharmaceuticals. To provide care to citizens during the future crisis, the city must demonstrate to the federal agency that it has a functioning plan. In turn, the CDC verifies that the preparedness plans of the community are compatible with the national infrastructure. The readiness of all parties is demonstrated through scenario enactment.

In the scenario created by CDC, the cities must distribute medical countermeasures to citizens after an attack that involves the rapid and widespread introduction of a virulent microbe into the population. The SNS has promised that it can deliver those fifteen truckloads of medical supplies anywhere in the lower forty-eight states within twelve hours. Thus the local governments are not tasked with acquiring, producing, or storing the supplies but rather with effectively distributing them to the population. Success is measured by connecting people with medication rather than by lives saved. The plan must target "circulations," enabling "good" things like pharmaceuticals to flow freely through the population while containing the "bad" things like illness.[54] The challenge, of course, is that the interactions that facilitate the positive flow of care are also likely to encourage person-to-person contact, which may increase disease transmission. A further challenge is to contain the disease without interrupting the economic and social exchanges that constitute the community. Rather than dramatically shifting social functions at the onset of a crisis, biosecurity operates by ensuring that the "normal" state of the community is one where viruses and disease have limited circulation.

Bioterrorism preparedness plans also transform existing circulations into mechanisms of distributing care during a crisis. Most CRI plans recruit the U.S. Postal Service to provide care by distributing the medicine. Postal workers will load trucks with the pharmaceuticals from a central distribution site and then deliver them along their regular routes, dropping drugs in every household mailbox. A 2008 study found that postal workers could deliver supplies to the population faster than a dedicated operation of emergency responders distributing them from a central site, thus demonstrating the value of using existing social systems during an emergency response.[55] Yet concerns over the safety of postal workers in this emergency role have led to decisions to pair them with armed guards, thus militarizing the routine social process of mail delivery.

This aspect of the preparedness plan assumes the at-risk population to be people dwelling in households that receive U.S. Postal Service deliveries. Planners also have to identify people who are part of the community but might be omitted from the postal lists, such as residents of nursing homes or prisons; people who might not be at their homes, such as hospital patients; and transitory members of the community, such as people staying in hotels. Many CRI plans call for the construction of temporary distribution centers where citizens can seek their own care, presuming they are aware of the event and can travel.

To test the components of its bioterrorism preparedness plan, the City of Albuquerque hired a contractor to design a scenario involving an aerosolized anthrax attack. Three tabletop simulations, in which planners tested three phases of the emergency response, were to be followed by a fourth, full-scale exercise. The cast included members of the public, particularly the individuals who would manage the community during the event. The scenario imagined a future in which security cannot be achieved without the consolidation of government around the singular goal of providing care, perhaps even through the militarized authority brought by armed guards escorting postal workers.

Albuquerque is the largest city in New Mexico, home to half of the state's one million residents. According to Greg Sanchez, the city's emergency director, the first task in developing the bioterrorism response plan was to define the population to be included in the plan.[56] The Albuquerque metropolitan area includes unincorporated county areas, one other city, an Air Force base, a university, and six Native American pueblos. Convincing those stakeholders of their shared vulnerability, let alone engaging them in crisis planning and response, was not a simple task. While some people effectively refused to participate simply by not returning phone calls, others produced more complicated narratives. The governor of one pueblo attributed his pueblo's nonparticipation to a cultural belief that articulating calamity invites it to happen. The population of Kirtland Air Force Base had to be included in planning because of its proximity to the city, but its residents are subject to military rule. Early drafts of the preparedness plan designated a warehouse on the base as one of three distribution sites, a decision supported by a Kirtland representative who was soon after transferred. His successor believed that were a terrorist attack to take place, the base would be put on lockdown, making the delivery of truckloads of pharmaceuticals onto the base highly unlikely. His opinion raised the question of demarcation, of whether the crisis would integrate residents of the base into the local security community or whether the military community would be cared for separately.

While nationwide scenarios may involve thousands of volunteers playing victims and responders, in smaller-scale enactments, much of the social effect comes from bringing together representatives of all the stakeholders and caregivers. At a July 2009 rehearsal in Albuquerque, a small auditorium was filled with uniformed police officers, city employees wearing polo shirts, representatives from Kirtland Air Force Base in military fatigues, and representatives from CDC dressed in business attire. After signing in and putting on name tags, the group sat down for an overview of the day's events. The contractor who designed and staged the mock events described various stand-in components of the exercise, such as rooms down the hall that would represent remote distribution points in the city. Then the curtain rose as the contractor played an audio recording of a simulated news broadcast, describing the "attack" and the ensuing public panic. The director then dispersed the group to their assigned stations, where they were connected to each other by phone and e-mail but could not physically see colleagues, who were supposedly at different sites across the city. All the participants acted according to their assigned roles within the Office of Emergency Management or respective government agencies.

While scenarios inherently depend on verisimilitude, participants are aware that the situation is simulated. Because there was no emergency, participants in this exercise glossed over or evaded difficulties that might have been significant in the field, such as transport time or not having accurate numbers on hand, while inventing new crises on the spur of the moment: "Hey, let's give them a gas leak over in building 3." Perhaps the important work being done here is not authentically reproducing the emotions of a crisis but rather delineating and building relationships. The simulation requires participants to assess resources, designate authority, allocate spaces, and build networks—acts that awaken a system of governance created for crisis management. The scenario work of the CRI rehearses a chain of care that links citizens to the federal government through the networks that distribute biological countermeasures. Here, preparedness is measured not by material possession of pharmaceuticals but by the creation of a plan that demonstrates how the local community will be subject to federal care during an emergency.

Scenario enactment enables governments to try out new ways of facilitating care during times of crisis and reinforcing the roles of citizens and government actors. Citizens learn to recognize face masks, white tents, and postal workers with police escorts as indicators of a bioterrorist event. It also allows the state to reminds individuals that the community is at risk of bioterrorist attack and establishes perpetual watchfulness and readiness as qualities of good citizenship. Such enactments also produce the images and narratives of security that signify preparedness, enabling citizens to evaluate readiness based on what they see in their communities. Good governance is enacted through the timely distribution of life-saving medical care after the event. The simulation not only affirms that the community is vulnerable but also enables the community to display its biosecurity practice.

LANDSCAPE TRANSFORMATIONS

Communities invest in biosecurity not only because they can see long-term benefits to the health and safety of their citizens but also because programs like the CRI, Operation Stonegarden, and the Urban Area Security Initiative spread funding dollars around the nation.[57] Simulating terrorism creates jobs, whether to design and plan the scenarios, to address vulnerabilities identified through scenario enactment, or to act out terrorism, as the residents of Playas do. An expansive government response to terrorism, supported by expertise created through scenario enactment, also increases the labor used to prepare the nation for biological threats. In communities where daily life has long been entwined with land and resource extraction, changes in the natural resource economy give rise to new ways to govern the landscape.

Remote and rural communities are part of the national security system. The vast landscapes of the American West have long created a landscape of fear in America's cultural imagination and continue to present a stage for the enactment of national security. European colonial expansion imagined a range of threats in the West, from Indians and bandits to drought and desert. The colonial project began through the conflicts that claimed territory and secured it against people and natures now classified as "Other." During the Cold War, every test of a nuclear weapon flashed a message of national security as it scorched the Western landscape. The nuclear program continues to inscribe colonial discourses into racialized landscapes in the West through uranium mining, milling, storage, and waste disposal that require native people to bear the radioactive burden of nuclear war in their bodies and natures.[58]

In the modern political state, Western natures are again integrated into the national security discourse, as in conversations about securing the inhospitable desert landscape against bodies crossing the U.S.-Mexico border. Security practices in the borderlands inscribe immigrants and the landscape in which they reside with biological risk. In this climate, the well-being of citizens in rural western towns like Hamilton, Montana, can be rationalized as an acceptable sacrifice for the survival of the nation. Current security practices, including scenario enactment, build a new infrastructure for the cultural imaginary of fear, creating another mechanism for sacrifice and dehumanized natures in the American West.

The modern security regime has made the borderlands of the Southwest into a zone where all life is suspect. Dozens of abandoned buildings dot the landscape, telling a story of communities that failed. They stand in stark contrast to shiny new security checkpoints, blimps, and watchtowers from which agents scan for foreign life in the landscape. Customs checkpoints break up miles of highway built to bring goods across the border. Here, federal agents with dogs on leashes scrutinize vehicles and cargo for suspicious organisms, whether a hidden human or a crate of fungus-ridden vegetables. The work of excluding bodies perceived to be dangerous, whether human or microbial, has created a new economic center for the borderlands, replacing resource extraction with national security.

Leaving Interstate 10 at Highway 113, one can drive south into New Mexico's Boot Heel, a wedge of land jutting abruptly into Mexico and then easing back to the Arizona border. This road ends forty miles from the border at an abandoned factory once known as La Estrella del Norte, or the North Star, because the lights on its smokestacks provided a beacon for immigrants crossing the U.S.-Mexico border at night. The abandoned smelter now speaks of the boom-or-bust economies of the West created by the glint of mineral wealth.

Built in the early 1970s by Phelps Dodge Corporation, the Hidalgo County smelter collected ore from the company's numerous mines to extract and refine copper locally before shipping it to market. The company built the smelter in the Boot Heel for proximity to mines in New Mexico and Arizona—two of the top three copper-exporting states—but also because the remote location separated population centers from the toxic byproducts of copper smelting.[59] With little attention to environmental woes, residents and political leaders in Hidalgo County rallied around the development, celebrating the influx of jobs and embracing a new identity for the area as an industrial center. As imagined in a 1974 comprehensive development plan for Hidalgo County,

> the smelter will make a substantial contribution to the economies of the immediate area and the State of New Mexico. In addition to the annual smelter payroll of around four million, Phelps-Dodge expects to pay nearly ten million dollars per year for services and supplies. . . . The increase of a significant number of industrial jobs in a community has a number of side effects. . . . In addition to an increase of the population to 900 there will be a rise in the average personal income. In the past, agriculture has been an important factor in the economy of Hidalgo County, but with Phelps-Dodge's expansion, this will change. The principal employment for the county will be mining and mineral processing.[60]

Composed while the paint was still drying on new homes for the smelter's employees, the fifty-five-page development plan fairly bursts with the anticipation of a new economic future. The plan connects the smelter to the quotidian experiences of Hidalgo County residents, conceptually binding the social existence of the county to the fates of the copper industry. Strong support for the smelter continued for decades, and a 1992 newspaper headline, "A Hidalgo County Dream That Came True," now framed on the walls of the Lordsburg Historical Museum, celebrated the completion of a million-dollar road-building project that would connect the company town and the county seat. Less than ten years later, that road would dead-end at an abandoned factory in the desert.

Nine miles from the end of the road sits a cluster of buildings, shade trees, and green grass. Designed in accordance with the suburban ideal, the company town of Playas plotted 250 two-and-a-half-bath homes on well-manicured lawns in neat cul-de-sacs. Banks, churches, bars, parks, a grocery store, and a bowling alley made this site a desert oasis, and children who grew up in Playas remember Fourth of July picnics at the pool, sneaking out with friends to drink beer in the desert, and breathtaking sunsets viewed from a neighbor's porch.[61] More than nine hundred people lived in Playas when the copper industry was booming, and in 1996, the Phelps-Dodge smelter was the second largest copper producer in the country, employing five hundred people and selling $488 million worth of copper.[62]

Changing technologies for smelting ore and the plummeting price of copper brought a swift end to the smelter. On June 30, 1999, the plant president warned Lordsburg's mayor, Arthur Smith, that "virtually all" Hidalgo employees would "sustain an employment loss between September 5–18, 1999."[63] The plant would keep a skeleton crew to close out operations, but eventually the entire town would be put up for sale.

Recognizing that the volatility of a mineral-based economy was once again striking the community, Hidalgo County planners revived the Hidalgo Area Development Corporation (HADECO), established in the 1960s to promote economic growth. HADECO was formed as a nonprofit group operating under the community-directed Hidalgo Medical Services (HMS), whose philosophy explicitly links citizens' health to their labor: "Healthy people are better workers, and productive workers are necessary for economic success."[64] Through HMS and HADECO, county planners promoted an image of the local economy quite distant from its land- and mineral-based roots, seeking to attract businesses to the Boot Heel through local infrastructure development and improved health services. Though framed in an argument about the unending fluctuations of agricultural and mineral economies and their accompanying transportation mechanisms, HADECO proposed to invest federal funds for health programs in ways that would boost local business development and attract new work. Thus, while Playas residents were closing up and abandoning their homes, a new planning philosophy was taking root, one that would reduce the community's dependence on natural resources while gaining federal seed money for a new type of development. At the time, the nation's attention was riveted on foreign terrorist threats and strengthening homeland security, and millions of dollars were being channeled into efforts to build the security infrastructure.

PLAYAS, NEW MEXICO: TERRORTOWN, USA

When Phelps-Dodge put the town of Playas up for sale, residents had the option to stay in their homes or include them in the sale. With the closure of the smelter and accompanying job losses, most people moved on. Public services closed down, and Playas seemed to be on the verge of becoming a modern-day ghost town. Then in 2004, with a $5 million grant from the Department of Homeland Security, the state engineering university, New Mexico Tech, purchased the entire town of Playas, planning to develop an antiterrorism training center. The town's remote location and empty suburbs offered the opportunity to rehearse for terrorism events out of public view, just as the nuclear-bomb designers of the Manhattan Project had retreated to the mountains of northern New Mexico seventy-five years earlier.

While the project was sold in Washington as meeting a national security training need, the local discourse focused on the promise of jobs in a struggling economy. At the ribbon-cutting ceremony in 2004, the president of New Mexico Tech, Daniel H. López, said, "I suspect, once we get going in Playas, the southwest region of New Mexico will have more economic activity and job opportunities being created for everyone involved than we can foresee at this time."[65] As with the smelter a generation earlier, this makeover of Playas was touted as an economic boon. Indeed, New Mexico Senators Pete Domenici and Tom Udall secured more than $100 million in federal and state funding to develop the Playas Training Center, which began with training for first responders and then expanded into military operations, including the construction of an "Afghan" village within the compound.[66]

In 2011, the center announced its first private contract, promising up to $27.5 million in revenue for New Mexico Tech.[67] Much of this income is invested in the facility itself, either for building new structures or for the upkeep and repair of the existing homes. Camera systems have been installed throughout the complex to record training exercises for later review; a new airstrip was built, with funding from the state, to accommodate large planes bringing trainees to the facility. These building projects send bursts of business to the community, but the training center is largely self-contained, with trainees accommodated on site and rarely venturing outside the facility.

The town itself is open to the public, though the guardhouse at center of the main road into town seems to indicate otherwise. "Tell the guard you're coming to bowl," instructs Anneliese Kvamme, from the center's public affairs department, "and you can come right in." The guard, however, discourages the casual tourist, politely informing visitors that the site is closed to the public. A sign on a chain-link fence barricading some of the training facilities parodies a popular cliché and echoes wartime

FIGURE 17. Training simulation at the Playas Research and Training Center, New Mexico. U.S. Air Force photo by Senior Airman Christina D. Ponte.

propaganda of earlier eras: "What happens in Playas, stays in Playas. Remember, information is power." Though residents boast that the bowling alley serves the best tacos in three states, the gatekeepers project a message that public use of this space compromises national security.

About twenty families live in Playas as employees of the training center. In addition to administrative work, the center provides two main types of jobs to locals: builders and actors. An actor might play the role of a suicide bomber cornered in a living room or a hostage in a standoff with a terrorist cell. Builders continually work to repair structures destroyed during scenarios. Such jobs exist because government and private entities believe that acting out emergency events in a place that looks like an American suburb, where performers detonate real explosives, will increase the safety of citizens throughout the country during times of crisis. The act of rehearsal, however, is itself a productive act, and the center at Playas exemplifies how simulation brings material change to the spaces where people live and work.

Though the scenarios at Playas are staged, the actors are also members of a community and dwellers in the landscape. Residents experience daily the sights and sounds of war, simulated or not, incorporating them into their world as readily as they greet guards on the entrance road or the daily arrival of helicopter transports bringing platoons of trainees. Locals attest that they can hear the explosions at Playas echoing for thirty miles across the flat desert. Still, they describe the work being done here as "patriotic" and "important to protect our country," saying they are glad to sacrifice the silence not just for the jobs but also for the security of the nation.[68] Thus, the desert continues to be a sacrifice zone, a place to simulate the next big crisis in a landscape that for generations has been deemed uninhabitable and uninhabited. It is a laboratory for violence, where technologies of government are remade by the biosocial experiment to know, anticipate, and prepare for future and foreign threats. Emergency planners play the roles of citizen soldiers, seeking threats in people's homes and neighborhoods and employing rapid, crisis-driven decision making to intervene, sometimes violently. The acts of war rehearsed in the New Mexico desert produce expertise about securing the American suburb and the type of citizenship it represents, and create a space where people can demonstrate that they are prepared to break down doors or quarantine neighborhoods for the sake of security.

Today, Playas's economy depends not on the value of minerals or the presence of copper in the ground but on the nationwide perception of risk and the financial support of a government investing in preparedness training. The terrorism simulation practices described in this chapter illustrate the effort to define how risk manifests itself in contemporary society. Federal and state governments have invested millions of dollars in southern New Mexico to imagine vulnerabilities and rehearse responses to terrorist threats. The political state has thereby shaped the cultural imaginary of terrorist threats, predicting what types of events are "likely" in order to justify the preparedness activities they rehearse. The political system accesses a limitless supply of imagined futures to rationalize the expansive politics of preparedness.

Success in Playas led to proposals to build similar training facilities in Alabama, Maryland, and Pennsylvania, and the proliferation of terrorism preparedness exercises from Congress to community meetings and classrooms shows the social acceptance of scenario enactment as a means of producing useful knowledge for a political response to terrorism. Because these centers have a specific geographical location and visible infrastructure, they stand as fixed reminders that the government is always engaged in preparing, even if the object of that preparedness is unknown. The message is that preparedness is not a one-time drill but requires daily rehearsal.

Beyond working to delineate potential threats, scenario enactments display the population's ability to respond to an incident, and not solely to assure citizens that they will survive and the nation is prepared to care for them during an event. Christian Erickson and Bethany Barratt argue that scenario enactments function simultaneously as terrorism deterrents, information warfare, and tools to manage public perception.[69] Large-scale simulations exhibit preparedness activities on a public stage, following the suggestions of the Graham Commission on Terrorism Preparedness that the most effective way to deter terrorism is to show potential terrorists that an attack on a well-prepared population will fail to create terror. This information warfare distributes the work of making war throughout the population.

Scenario enactment evokes the war games and virtual warfare that James der Derian described as the "virtual continuation of war by other means," where the media and technologies of contemporary warfare sustain the military-industrial complex.[70] Enactment becomes a technology to extend warfare into the domestic arena, bringing the sights, sounds, and smells of a terrorist event into people's homes and communities. Whether or not such simulations deter attacks, they teach citizens how to identify an act of war, making common the unfamiliar materialities of war. Enactments invite communities to organize around a crisis, rehearse systems of power and authority, and, in large-scale events, literally act out the effects of a nuclear blast or plague virus on their bodies. In these acts of violence, citizens study war and learn how to militarize their own communities in a crisis.[71]

New forms of national security rely on this never-ending state of war to institute and maintain social systems. Enactments create a future where violence is expected, rationalizing the mobilization of citizen soldiers to protect the systems that sustain the life of their community. Microscopic organisms—materialized as weapons, built from DNA sequences, and genetically modified for pathogenic superpowers—create a potent natural landscape for threats, demanding new institutions to care for citizens. The connection of germs to terror imagines threats that rapidly evolve and move through a modern global landscape in ways that put every human body at risk, as evidenced by the lines of air travel that painted the entire globe with infectious disease on the model I played with at the museum. Masco claims that "hypersecurity has become a dominant mode of governmentality after 9/11, a series of linked discourses and official practices that work through the mobilization of a named or unnamed, but always totalizing, threat."[72] The modern preparedness industry relies on the sustained perception of a threat that, even if not totalizing, is deeply personal and targeted: through infection, a germ attack is personal, and through contagion, it contains the potential to be totalizing.

Enacting a fearful future constitutes a profoundly political act of making meaning. Acts of governance rationalized through modeling and simulation have high a potential for creating expert knowledge. This expertise disperses through the population, making a range of citizen actors into experts in judging how differences in human behavior indicate threat or vulnerability. When the nation is continually at risk of terrorist attack, particularly unseen biological events that may already be under way, the responsibility to intervene is ever-present. Deviant bodies and behaviors may be perceived as vulnerabilities in any community, and political acts may target them, whether through the rule of law or social movements. Furthermore, the many unnamed threats of the war on terror can potentially be linked to any individual or community. The National Asset Database, which registers potential terrorist targets, was ridiculed in the media in 2006 for identifying local festivals, petting zoos, and ice-cream parlors as potential terrorist targets, but the catalog of nearly eighty thousand sites shows the flexibility of terrorism discourses in connecting the terrorist threat to every citizen and every community in the nation.[73]

Before they acted out a terrorist attack in TOPOFF3, participants in the Concord, Connecticut, drill gathered for a community dinner. Rituals of food and camaraderie offered a stark contrast to the fracturing of social space created during the attack. Similarly, the quiet of the desert and the pacing of rural life is disrupted daily in Playas, New Mexico. This contrast is key to the social effect of the exercise, showing what is at stake in an attack. Transformation into a crisis-response mode can be seamless when the security apparatus is in place. Whether for a few hours during a national-level exercise or on a permanent basis through the wholesale purchase of a town, simulations set a precedent for the conscription of a community into the nation's defense complex. This work stabilizes security practices at the local level, integrating them into local economies and governing practices.

When communities rehearse a common response to terrorism, they affirm their shared vulnerability to foreign threats. New Mexico's location on the U.S. border has driven the state's involvement in national security practices. In a place where the regulation of bodies across the border is already scrutinized with aggressive watchfulness, there is a precedent for growing a security economy. The continuing war reverberates here, in the state where the detonation of the first nuclear bomb created new natures and landscapes that could only be understood through a history of violence. The security infrastructure on the U.S.-Mexico border defines the nation in terms of biological risk, bringing together scenario enactment, microbial science, and security technologies to secure the borders of the nation-state.

6

Bioterror Borderlands

Of Nature and Nation

New Mexico Highway 9 hems one hundred miles of United States soil on the southern border of the nation, a river of pavement meandering through the desert where the most visible markers of nationhood are five-mile stretches of border fence and green-striped Border Patrol trucks. Locals claim that if your car breaks down along the isolated stretches between the towns of Hatchita, Columbus, and Sunland Park, in less than fifteen minutes an officer will pull up to lend a hand. Agents approach anyone who lingers, suspicious that they may be waiting for illegal travelers crossing the desert from Mexico on foot. Driving Highway 9 on an afternoon in July 2009, I passed one sedan with local plates and twenty-three Border Patrol vehicles, including a Chevrolet Suburban inching along the shoulder of the road while the driver hung his head out the window, scanning for human footprints crossing a smooth path raked in the sand near the highway. White blimps hovering above the road heighten the sense that every movement along this border is watched. After traveling through this landscape for an hour, I developed a keen sense that my presence was unwelcome and suspicious.

Although the story of the militarization of the U.S.-Mexico border begins and perhaps ends with the question of immigration, I had come to the borderlands to explore how bioterrorism preparedness work on the nation's borders sustains notions of nationhood and citizenship. The caravan, the fence, and the blimps are evidence of a booming security economy in the region, drawing federal funding to support jobs in construction, enforcement, and expertise. Less visible to the highway traveler are the technologies that regulate the movement of life through the borderlands, seeking to create a biological buffer zone. As life forms move, seen and unseen, through the landscape, they engage with multiple others. Many disciplines study migration with the hope of explaining mass rearrangements of life on the planet. In biology, migrations are described as seasonal and patterned to procure food; in the humanities, migration means movement with the intent to settle. Both these types of migration, and many others, are subject to the mechanisms of the security state that aim to command and control life itself.

To protect their territory and police their borders, governments build systems to regulate intentional and unintentional movements across borders and between populations. The goal of security is to limit the movement of "bad" objects while still enabling the flow of goods and services that sustain the nation. To create biosecurity, governments work to duplicate on a large scale the function of laboratory biosafety cabinets: they must control the flows that would bring potentially harmful life into contact with the world in which citizens live and work. Microbes cling to other forms of life, including organic matter, such as fruits, vegetables, and human and animal bodies. Governments must diligently watch for micro-hitchhikers that might harm their citizens or damage the food supply. Like the "coyotes" who smuggle migrants across borders, livestock, produce, wildlife, air, and humans transport microscopic life across borders. In the biological borderlands, all movements are suspect for the latent threats they animate.

The recognition of the U.S. border as a fluid space, crossed and inhabited by biological entities of all sorts, informs how border residents undertake national-security projects. A fence is not a biological border, but the daily practices of residents and workers can define the natures that belong to the nation and those that must be excluded. Here governance must promote the circulation of desired goods while attempting to exclude unwanted or foreign bodies, large and small. By aligning biological movements in the borderlands with terrorism and disease risk, national security naturalizes a politics of exclusion. The decisions about whom and what to turn away are codified by law but enacted in the field, turning the production of biosecurity over to the planners and patrollers who work in the borderlands.

Among the many natures that circulate across borders, microbes of all sorts, and particularly infectious agents, define boundaries of nationhood and citizenship. Disease reorders geopolitics, for it moves by the associations of bodies, defying borders at the same time as it "provokes their fervent reaffirmation," in the words of Priscilla Wald.[1] Modern systems can move disease around the globe in moments, but rather than collapsing national boundaries, this movement has promoted the fervent defense of the nation-state against outside threats. Whether describing Africa as the biological source of all emerging infectious disease or quarantining immigrants, the nation works to protect its own citizens from outside threats. These cultural materializations of our fears of disease are practices that, using biology, define who we are in opposition to others.

Biological boundaries, like all borders, are fluid, and when they move, they shift the balance of power. Brian Massumi argues that boundaries are less limits than thresholds, and that they are actually set through the act of passage: "The crossing actualizes the boundary—rather than the boundary defining something inside by its inability to cross. There is no inside, and no outside. There is no transgression. Only a field of exteriority, a network of more or less regulated passages across thresholds."[2] A microbe that crosses a boundary between nations does not know it has committed a political act, but in the crossing, the microbe has potentially changed the calculus of its own existence, the political state of host bodies, and the security of a nation. Thus, through crossings, the borders of the state are repeatedly actualized by the flow of goods, violence, and people—objects that take on new meaning as they move through a political landscape.

This chapter illustrates key themes of the book through three studies of biopolitics in the borderlands, showing how the modern security state is naturalized through the work of managing human and nonhuman life in the desert Southwest. The first case is historical and exemplifies how military might blends with biological control in response to fears of disease. Events on the border in the 1910s naturalized social fears of race through the inspection and policing of immigrant bodies, coinciding with military shows of nationalism. Next, I offer a snapshot of how the United States polices food transportation across the border to inform a discussion about the difficulties of multistate and multinational terrorism-preparedness planning. Finally, I briefly consider recent and recurring outbreaks of pandemic disease as opportunities for the nation-state to assert authority over its borders and its citizens' bare life, asserting the expectations of citizenship in regard to policing health and risk in a continual state of emergency. Borderlands offer an opportune site for examining the tensions between security and terrorism, life and risk, human and nonhuman, that characterize modern governance.

Through border security, bioterrorism preparedness is reconstituting the geographies of the nation, including its relationships to its own citizens and to other nations. Policing national borders to exclude biological threat defines the limits of the nation in terms of people's interactions with geopolitical environments. Making nature in the borderlands is part of the work to make the nation after 9/11.

COLUMBUS: NATIONAL SECURITY, 1916

Residents of Columbus, New Mexico, have taken the idea of the snowbird—an individual who relocates to warmer climates during the winter—to its extreme, driving south with their homes in tow until they are within spitting distance of the Mexican border before dropping the trailer hitch and pulling out the barbecue grill. From a south-facing porch, the snowbirds can look across the border to the Chihuahuan town of Puerto Palomas. A border-crossing station sits between the two towns, but residents regularly commute between them. The Americans head south in search of cheap prescriptions and dentist's appointments; Mexicans come north to attend school and to fill jugs with clean, free water from the spout in front of the Columbus civic offices. The fountain is a generations-old tradition symbolizing the truth that although these two communities are increasingly rent by fences and politics, neither town would exist without the other. Their fates are linked by blood, water, and money.

More than a century ago, Columbus was already bustling with border-protection business. Camp Furlong, on the outskirts of town, was one of several army camps displaying military might on the nation's border during the revolution in Mexico. In the days following Pancho Villa's famous 1916 raid on the border town, the arrival of additional troops swelled Columbus's population to more than fifteen thousand, making it briefly the largest city in New Mexico. On the night of March 9, 1916, the Mexican rebel leader rode into Columbus with a small army of followers. The rebels looted and burned the town in a surprise attack that lasted more than an hour before soldiers and machine guns drove them out. At least seventy of Villa's band died in the raid, along with eighteen U.S. soldiers and citizens. From Columbus, the U.S. army pursued Villa far into Mexican territory, employing automobiles and aircraft in combat for the first time in U.S. history. Aircraft searched for the Villistas from above, and pack horses carried fuel for armored supply vehicles to support the army's march from the north. Some claim the army's year-long pursuit of Villa and the Border Campaign played a vital role in preparing U.S. troops for entry into World War I: such rhetoric identifies Mexico as little more than a training ground for U.S. military might.[3] Thirty thousand soldiers participated in the pursuit of the Mexican bandit, and by mid-June more than one hundred thousand members of the National Guard were stationed along the border to handle other skirmishes incited by the Mexican revolution.

The motivations for Villa's raid on Columbus remain in dispute. It has been read as an act of revenge for a bad weapons deal or for America's alliances in the Mexican revolution. The preceding years were marked by racial segregation and violent conflict in the region. They were also characterized, however, by progressive health reforms. General John J. Pershing's troops spent May 1915 "cleansing" a Mexican district in El Paso referred to as the city's "plague spot." Pershing's troops hosed down streets, burned refuse, and demolished dwellings as part of an initiative to "improve" the district. Meanwhile Villa showed up in newspapers as a character in complex narratives linking disease and violence with Mexican people. Several days before Villa rode into Columbus, a fire in a jail in El Paso that killed about twenty Mexicans was linked both to Villa and to delousing practices intended to combat typhus. The kerosene used for the delousing ignited, and the resulting blaze swept through the jail. Rumors circulated that Villa had promised to "make torches" and "set people aflame" in revenge.[4]

Pershing's failure to apprehend Pancho Villa after a ten-month campaign brought uncertainty to the borderlands, rationalizing a surge in public health controls, nominally for the prevention of typhus. These measures worked alongside military action to manage bodies crossing the border, but through different channels. Alexandra Minna Stern relates a gripping tale of nation building through public health at the Ciudad Juarez–El Paso border crossing. On January 23, 1917, the U.S. Public Health Service enacted an "iron-clad quarantine" on all bodies entering the United States from Mexico.[5] The announcement was immediately followed by riots, reminding us that power never flows only one way and that groups protest governance that defines them as infective.[6]

Stern points out that the immigration inspection procedures at the El Paso crossing differed in intimate effect from those at Ellis Island in New York and Angel Island in San Francisco during the same period. In those ports, medical inspections were handled by ship companies: only the diseased were stripped and scrutinized, and delousing was uncommon. In El Paso, all individuals were separated by sex and forced to strip naked. Their clothing was taken to the laundry for a half hour of chemical cleaning while they were scrutinized for lice. If men had lice, their hair was shaved off and burned; women had a treatment of kerosene and vinegar applied to their heads and hair for half an hour. Following the delousing, people were required to shower with soap and kerosene and then vaccinated against smallpox. The inspection concluded with an examination for physical or mental defects that included psychological profiling and questioning about citizenship. This dehumanizing scrutiny continued into the 1920s. The facility in El Paso inspected about 2,800 bodies per day, or about eighty per hour by each of three working physicians.[7]

Stern eloquently characterizes the health inspections on the U.S.-Mexico border as work that brought empire, the body, and "biologized techniques of social differentiation" to bear on the rise of U.S. modernity. She writes of boundaries that both embrace and repulse: "Boundaries, at the edge of the empire-nation, moved reversibly from the epidermis or body itself, to the landscape of rivers and deserts, and onto bodies en masse, or 'races,' as classified by censuses and other indexical strategies."[8]

As the government presence on the border grew through the creation of bureaus of immigration and naturalization, along with the U.S. Border Patrol, the border was further medicalized, militarized, and racialized. Even "locals" were required to pass regularly through the disinfecting station, receiving a certificate of inspection that expired after a week. Border crossings became suspect and required the scrutiny of physicians to declare racialized bodies to be clean. Microbes adhering to Mexican bodies defined the quarantine experience. Stern quotes one woman's recollection of the purpose of the disinfecting station: "They thought [Mexicans] were bringing microbes or something like that over from Mexico."[9]

Significantly, the public health search for microbes continued long after the abatement of the typhus epidemic. This act of medical control, instituted during a state of emergency, became part of the quotidian experience that made the border. The politics of the disinfection station extend deep into the nation, materialized in bodies on both sides of the border through the public display of health and cleanliness as rites of citizenship.

After Pershing's failed campaign, as the unrest in Mexico diminished and conflict increased overseas, the military relinquished the groundwork of border protection to nonmilitary entities. Until the Border Patrol was established in 1924, the border was delineated largely by the public health mechanisms that inspected migrant bodies. When Pershing and the soldiers left Camp Furlong, the population of Columbus dwindled. When the railroad left town in the 1950s, mapmakers began to mark Columbus as a ghost town.

In the twenty-first century, drug violence south of the border—including beheadings, the murder of the mayor of Palomas, and a much-publicized incident in which a dentist was robbed at gunpoint while the mayor of Columbus sat immobilized in the dentist's chair—has slowed the demand for pharmaceuticals and dental work on the Mexican side of the border. Violent shifts continually remake the border. One Columbus snowbird assures me that she feels perfectly safe in town, for once the smugglers cross into the United States, they flee "like a prison break," far from the border itself.[10] Palomas, however, suffers under the violence, and many residents migrate, some taking up permanent residence in its sister city to the north. Tourist attractions south of the border sit empty.

In this time of unease at the border, the annual Camp Furlong Day, a commemoration of the 1916 attacks, seems perhaps more significant through its effort to make sense of the biopolitical struggles in the borderlands. Formerly called Raid Days, the festivities center on a reenactment of Pancho Villa's attack on Columbus. Led by an actor portraying the famous bandit himself, a cavalcade of Mexican riders travels over the border, where they meet a group of Americans also on horseback. Rather than opening fire, the two groups parade together through town in a show of reconciliation. The festival draws celebrants from both sides of the border, even as naysayers call it an inappropriate celebration of the most significant terrorist attack on U.S. soil prior to September 11, 2001. One *USA Today* reporter likened the town's Pancho Villa State Park to an imagined Osama Bin Laden State Park in Lower Manhattan, where tourists "learn how the Muslim militant planned the attack on the World Trade Center and the Pentagon, how he eluded U.S. forces, and why he hated America."[11] Indeed, the creaky town museum, crowded with collectibles of the Pancho Villa days, describes the events of 1916 and the actors on both sides: Pancho Villa the bandit and Pershing the war hero. One Columbus resident, pointing out a likeness between the thick-mustached Pancho Villa and Saddam Hussein, whispered that Villa "just looks like he was a terrorist."[12]

Because the name Pancho Villa carries the weight of legends and a heap of Hollywood glitz, locals hesitate to give up the famous name and its tourist draw. Rather, as the Pancho Villa State Park manager argued, they use the name and celebration to honor long-standing bonds between the two communities and "the celebration of our common heritage."[13] Residents of the borderlands claim identities that are neither fixed nor homogeneous and embrace a complicated narrative of nationhood, conquest, and economic dependency. Though the technologies that create the U.S.-Mexico border have changed over time, both the annual parade of Pancho Villa on U.S. soil and the daily parade of individuals crossing the border in pursuit of water and health care are reminders that there exists no stable subject in national security discourses. The changing security economy produces new ideas of citizenship and value for residents of the borderlands.

SANTA TERESA: BOVINE BORDERS

About a mile down the border from the port of entry where the modern-day Pancho Villa crosses into the United States with his *cabalgata* lies a facility where thousands of four-legged Mexican immigrants are scrutinized before being granted entry into the United States. Mexico exports almost a million cows to the United States each year, supplying the large herds of cattle that graze in Texas, California, and the Midwest. About a third of these animals pass through one of the two ports of entry in New Mexico, where they are held and inspected for signs of disease.[14] The work of these checkpoints is to prevent diseases such as tuberculosis from being transmitted to U.S. cattle and decimating a multibillion-dollar beef industry. Here, distinct qualities of nationhood are negotiated in bovine bodies: animals that manifest health are permitted to cross the border, and animals displaying infection are sent away.

This work is done in the name of protecting the U.S. food system and citizens' bodies, but it must also be done in a way that ensures the supply of vital resources to U.S. markets. The exchange of goods, including food and livestock, makes the border a dynamic space where moving bodies are scrutinized and officials attempt to decide which life forms can move where. Biology overlaps with nationality when federal agents mark a body as suspect, and turning away a body from the border because it is biologically risky is a mechanism of national security. Attacks on ranchers and agriculture inspectors in the borderlands have drawn attention to the work the government does here to regulate the agriculture industry, entwining discussions of ranching, terrorism, immigration, and drug trafficking and creating a complex representation of cows and germs as national security threats.

The Santa Teresa–Jeronimo cattle crossing is known as an efficient and effective inspection facility where cattle literally walk across the border, in contrast to sites that require animals to be moved by truck. When they arrive from ranches in Mexico, the cows are held in quarantine for fourteen days, allowing time for signs of diseases to manifest themselves. Before the cattle cross the border, veterinarians test for infection, and cow handlers run the animals through disinfecting baths before prodding them northward to the U.S. side, where they are tested again. Dozens of semi trucks with a rainbow of license plates from U.S. states line up outside the facility, ready to transport the animals into the flows of the livestock industry. Union Ganadera Regional de Chihuahua, a Mexican livestock cooperative working on behalf of more than three thousand ranchers in northern Mexico, runs both sides of the Santa Teresa facility and the Mexican side of a similar facility near Columbus. To ensure that all human bodies can do their work but still end up on their assigned side of the border at the end of the day, the handlers wear vests, allowing border patrol officials to watch and regulate the workers' bodies even as the workers manage the bodies of cows. The daily operations here constitute the border as a space where both biology and nationality are suspect, and bodies of all sorts must be policed.

Like other forms of border security, agricultural inspection practices have been infused with new security measures following drug-related violence near the border. In 2010, agriculture inspection work that was once done on the Mexican side of the Texan border was relocated to the United States after inspectors from the U.S. Department of Agriculture (USDA) were held at gunpoint at the Nuevo Laredo inspection facility. FEMA funding, provided through the border-security initiative Operation Stonegarden, provides armed escorts for New Mexico Department of Agriculture (NMDA) workers who perform annual inspections of livestock scales on the remote ranches in the border region. This measure is deemed necessary because almost all actions are suspicious this far from paved roads and cell phone service, and the distinction between a Border Patrol agent and a weights-and-scales inspector may be unimportant to an armed and cornered participant in illegal border activities. As in the Cities Readiness Initiative, which has postal workers distributing vaccine accompanied by soldiers, the threat of violence is once again met with armed escorts. The routine work of government to regulate the beef industry becomes a display of military might and a territorial claim that forcibly asserts the power of law in the region.

After the much-publicized murder of an Arizona rancher, Robert Krentz, around the same time, Governor Bill Richardson deployed the New Mexico National Guard to the borderlands in an executive decision reminiscent of presidential directives after the Pancho Villa raids nearly a century earlier. These state-initiated acts accompany a broader taking up of arms by residents and citizen militias in the area. One local gun shop reported a 20 percent increase in sales following the Krentz murder.[15] Both government and citizens are engaged in the production of violence in the borderlands, a militarization emerging from a belief that the rule of law—whether immigration law or the fair and accurate calibration of livestock scales—must prevail.

The militarization of daily life on the U.S.-Mexico border links the uncertainty of individual survival to the broader economic concerns of the state, a fusion of purpose that is increasingly evident in contemporary forms of terrorism. Arjun Appadurai argues that 9/11 bound local histories and political turmoil in a new "geography of anger" and "freshly charged [the] relationship between uncertainty in ordinary life and insecurity in the affairs of states."[16] Uncertainty pervades the borderlands, whether about the presence of disease or an individual's right to be present on one side of the border or the other. The disruption of certainty produces anger, and anger breeds violence. The government's security response must simultaneously attend to the well-being of individual citizens, the protection of government workers and their labor, safeguarding livestock and livelihood, and the preservation of a diplomatically guided economic relationship between nations—in the name of fighting drug smuggling, illegal immigration, terrorism, and disease.

LOS CRUCES: "PESTS DON'T KNOW BORDERS"

While protecting the state's $6 billion agriculture industry provides a local incentive for national agricultural security, New Mexico's borderlands are also a buffer zone for an ecological boundary that cannot be policed by fences or other technologies.[17] Not only do mountain lions and butterflies move through the landscape without regard to national borders, but people, cattle, cotton, and bananas also carry microbial hitchhikers across borders, potentially bringing disease into the nation and the nation's food system. The NMDA's Southwest Border Food Safety and Defense Center (SWBFSDC) was established at New Mexico State University in 2005 to grapple with this particular problem.[18] Its director, Billie Dictson, is quick to point out that eating in the United

States depends on the safe transportation of edibles across borders.[19] Dictson claims that the agriculture economy employs 17 percent of the American population, not counting consumers. Contamination in the food supply threatens lives: more than five thousand people die every year from food poisoning.

In reports that read like major drug busts, border patrol agents boast of "seizing and destroying" hundreds of pounds of illegal foods, laying wrapped meats and packages of produce on the evidence table like bricks of cocaine.[20] These tokens are offered up as evidence that security has done its work to protect against the accidental or intentional introduction of contaminants into the food supply, thereby preserving a vital system that sustains citizens' lives nationwide. Preserving these systems is the primary goal of the Southwest Border Food Safety and Defense Center and its government partners. In a 2007 brief to Congress, the Department of Homeland Security promoted nongovernmental initiatives like the SWBFSDC as a way to "solve security and expedited movement of people and goods across our borders." Even if these partnerships focus on natural disasters rather than terrorism, DHS argues that the combination of protection and preparedness actions produces "transferable" results.[21] Preparedness for the naturally occurring threat is described as promoting readiness for the intentional, and vice versa.

The objective of these biosecurity measures is to contain biological threats without hindering the vital flow of animals and vegetables through the borderlands.[22] Following Foucault, Nick Bingham and Steve Hinchliffe suggest that regulating circulations in this way is a key characteristic of biosecurity.[23] Biosecurity practices attend to the specific qualities of environments that favor such circulations. It is not so much concerned with dramatically rebuilding spaces as with modifying existing spaces to maximize the circulation of "good" things. While recognizing that the only way to guarantee that no bovine disease would cross the border would be to stop the imports entirely, biosecurity practices focus on removing the specific threats by using a facility to identify and remove unhealthy animals. Similarly, the Border Food Safety and Defense Center seeks intimate knowledge of the biological and cultural conditions of the borderlands in order to manage preparedness programs for the maximum benefit of the area's agrarian economy.

While the statement that disease knows no borders may seem cliché, addressing biosecurity in a diverse and sparsely populated border region poses serious challenges. Dictson claims the biggest challenge to his work to secure the biological border has been putting disease "on the radar."

Thanks to H1N1 (swine) flu, Ebola, and *E. coli*–contaminated spinach, people are thinking about disease as never before, and Dictson says the task now is to "take them from the awareness level and get them to a place to respond."[24] Not only are people confused about whom to call during agricultural emergencies, but in this region responders are often called on to think creatively about, say, how to round up one hundred head of escaped cattle. Dictson claims that people are very good at handling logistics in daily life, but they don't know how to apply those skills to organized emergency response. He identifies a suite of skills, similar to those used by CDC employees or targeted in large-scale scenario enactments, that must be brought to bear on a biological event. Here again is a system of governance that is continually building and rehearsing an emergency response to attain its objective to secure the borders.

As part of an international agro-security project, Jeff Witte runs the Ten States Border Training Initiative, involving participants from all states on the U.S.-Mexico border in planning and rehearsing a coordinated response to a food-systems emergency in the border region. Witte staged three trainings from 2004 to 2009, including tabletop exercises and a full-scale simulation. These exercises involved more than 250 people, from governors and Homeland Security agents to USDA inspectors and veterinarians. Participants role-played responses to events such as a foot-and-mouth disease outbreak and an apple moth infestation. The dual acts of planning and performing emergency events not only establish mechanisms of response but also prescribe roles for two national governments in caring for their citizens' health and economic prosperity.

Developing a scripted and sustainable preparedness system with two federal, ten state, and numerous local governments is a particular challenge for the Border States Training Program. For example, Dictson observes that whereas U.S. agencies fight for local control and readily act within their perceived jurisdictions, the Mexican states want federal government sanction before proceeding with local actions, causing delays.

Another problem stems from the different languages used—not just Spanish and English but also the acronym-laden government jargon. Witte has been working to ratify the use of the favored model for emergency response in the United States, the Incident Command System (ICS), as part of the border security plan. He recently received a call from a Mexican planner who said he had finally received federal permission to use the "CSI" model, confusing the correct acronym with one made famous by a television series on crime scene investigation. Emergency response assumes a shared language and purpose, and scenario enactment works to establish common ground between groups that may not share such a focus in their "normal" routines. For Witte, scenario enactment centers on overcoming difference and prescribing a common purpose for the participants while giving them a working vocabulary and other tools for imagining a common threat and a shared future.

Preparedness exercises with an international scope, such as Witte's program or later NLE scenarios, also work to export U.S. models of governance. The ICS lays out a structure of authority, scope, and accountability that is designed to be applied to any "incident," defined as an unplanned situation requiring a response. Originating in the 1970s among firefighting agencies, ICS is designed to be scaled to an incident of any magnitude by adding layers of command structure as more people "enter" the incident response. It was also created to enable multiple agencies of government to work together, minimizing miscommunications and duplication of effort. ICS operates under the assumption that every incident has an end point, when the structures of authority implemented during the event are eliminated and people return to their "normal" place in the social structure. But the DHS goal of creating a "national approach" that "treats crisis management and consequence management as a single, integrated function" blurs the boundary between incident and outcome.[25]

In 2003, Homeland Security Presidential Directive 5 required all federal agencies to train their employees in ICS, and all state and local agencies working with federal funding to use ICS in their preparedness planning.[26] By controlling the purse strings, the law extended the reach of ICS deep into the public and private sector, recruiting participation from groups like the Border Food Safety and Defense Center by establishing the terms of their access to federal funds. The law requires citizens to rehearse their response, thereby institutionalizing the structure and mechanisms of incident command within existing forms of governance. Not only does this create potential for systems of government to shift at a moment's notice in the event of an incident, but it instills in the population the notion that the state of exception is nothing more than an extension of normal institutions.

The system's name originates from the idea that the first person on the scene of an incident assumes authority over the response from the moment of their arrival until the incident ends or command is transferred to another individual. ICS presumes a sole authority over the incident, with a strict hierarchical structure and fixed roles (though different people may fill those roles at different times). The system is reactive, structured for responding to crises rather than planning how to prevent them.

ICS presents a streamlined system of governance that can be rehearsed both through simulation and by application in a range of incidents, from forest fires to hurricanes. In training modules individuals learn that working outside their assigned roles leads to negative consequences, such as slowing the response, wasting resources, or instilling panic.[27] ICS promises that establishing hierarchical authority will create "the capability to work efficiently and effectively together."[28] Regular rehearsal of this response through preparedness exercises further instills familiarity and proficiency in new ways of managing citizens.

Adoption of the ICS is a requirement for partner agencies to get federal funding. By training emergency responders of all sorts, including private entities and individual citizens, in the ICS, the U.S. government teaches people about systems of authority and resource management, along with shared terminology and expectations, practices meant to be put into play at a moment's notice. Through international collaborations for preparedness planning, U.S. agencies export the ICS abroad. Countries that wish to collaborate with the United States to command human and nonhuman enemies are required to adopt the preferred U.S. system of control. The United Nations has promoted the international use of ICS, apparently anticipating that diverse systems of government around the globe could be suspended in favor of a unified approach during an incident on a global scale.

Witte and Dictson give the impression that the biosecurity business is booming. They are undertaking a wide range of projects, from helping New Mexico communities develop terrorism-preparedness plans to distributing DHS brochures and twenty thousand emergency-response bags to elementary schools. A leader in border preparedness, the group has received a $2 million grant from DHS to develop a national curriculum for multiday courses on preparing communities for agroterrorism. DHS has stringent curriculum requirements, and the proposed textbook-length manual will undergo peer review at the national level and mandatory updates every three years.

A course with the DHS seal of approval can be accessed for free by communities who receive DHS funding, saving the expense of contracting with a private company for biosecurity training (between five thousand and half a million dollars). However, DHS stipulates that its funds cannot be used to train people who are foreign nationals, limiting the ability of responders to train together for a unified multinational response. Dictson is concerned about this restriction because it is frontline agricultural workers who will see disease first and are therefore the ones who need to learn how to recognize it. He reiterates that "pests don't know borders" and that Americans will benefit from educating individuals working on both sides of the border regardless of citizenship. Dictson believes the culture of secrecy on the border, along with a tradition of covering things up and working independently, creates a false perception that locals won't work to secure the food system, but "though training, awareness, and scaring the hell out of 'em, now they just might call someone when they see that white powder."[29]

Preparedness work in the U.S. borderlands racializes the production of biosecurity. On both sides of the border, the landscape is populated by Mexicans: their bodies are part of the biological systems of the region. Dictson's work points explicitly to a racial gap in the U.S. preparedness strategy, whereby populations are excluded from the national preparedness program because of their nationality, and he works to extend the civic practice of preparedness planning beyond white suburban populations. Agricultural workers in fields around the globe may be the first to spot a biological threat in the food system. The work to move food across borders similarly involves low-wage labor and jobs performed by people of color. If securing the nation's biological systems is the objective of governance, these populations must be brought into national security practice. The lack of attention to the diversity of the population in preparedness planning suggests that it centers on promoting a particular type of governance in the present, in which planning for the future means creating and securing a uniform suburban citizenship for people today.

The national security discourse presumes a homogeneous population whose response to any incident in any location will be structurally the same. Difference, including racial difference, may be more closely aligned with the threat than with the response. Discrimination and cultural fear of people from the Middle East since 9/11 famously exemplifies the racialization of national security. The security practices that have become the background noise of daily life in the United States reiterate the alignment of cultural and biological difference with threat, reminding citizens that their very bodies carry the risk of deviance.

Biosecurity work in the borderlands not only secures that region against threats to local agriculture but also produces security for the entire United States. The border becomes a place of sorting, where suspect organic matter is denied the opportunity to intermingle with the biological ecosystem inside the national boundary. The cultural formation of a border therefore constitutes the biological space within. Although the permeable border certainly does not prevent all foreign organisms from crossing, the work of policing it affirms the idea that a nation can determine its biological landscape, denying the entry of living organisms, both human and nonhuman, in order to achieve its security objectives and create the natures it desires.

MEXICO: SWINE FLU, 2009

Just as flu follows seasonal cycles, the post-9/11 security state has accessed the social effects of pandemics in regular cycles that keep disease present in the public mind. SARS, avian flu, swine flu, Ebola virus, and Zika virus have appeared on the global stage just a few years apart, reviving fears of nature and demanding governance. Though politically different in effect from bioterrorism, these events play out through the institutions and relationships created for bioterror deterrence: public health workers, government agencies, community organizations, and so on. These events also create opportunities for the nation to demonstrate preparedness. Whereas controlling hurricanes or earthquakes seems inconceivable, microbes represent a form of nature so intimate and personal that they appear to be within the bounds of control, and people can and do work to regulate them within their own bodies and environments. The inability of a nation to contain its microbes might thus be understood as the failure of a government to manage its citizens.

In this last section I examine the national and international response to H1N1 (swine) flu in 2009 to explore how pandemic events constitute particular forms of nationalism in the security state. Germs so readily cross borders in the modern age that boundary crossing has become a moment that signals a threat. Infectious disease enters our awareness particularly through the crossing of borders, as people watch news reports to track pathogens that might move closer to their homes. Although the threat of pandemics differs from the threat of terrorism, the ontological grouping of these threats has mapped a terrain for disease and fear that has consequences in the production of citizenship in the national security state.[30]

Take, for example, the geographical labels of strains of flu: Spanish influenza in 1918, Asian flu in 1957–58, and Fujian flu in 2003–4. Epidemiologists claim that the labels have medical value in identifying where an outbreak originated.[31] At the same time, they acknowledge the potential of such naming to harm people's perception of nations. When China did not report an outbreak of SARS for three months in 2002–3, the international public health community blamed the nation for the severity and spread of the disease. By contrast, Mexico was lauded when in 2009 it announced the outbreak of a deadly flu virus just ten days after health authorities identified a possible disease emergency, but this announcement incurred significant costs. While Mexican citizens suffered from the disease, the Mexican economy suffered from declining trade and tourism, losing an estimated 1 percent of its gross domestic product.[32] With news of an unusually deadly virus in the Northern Hemisphere, nations around the world suspiciously eyed their borders (and people who crossed them) and prepared to draw up the barricades to keep out the disease.

The first H1N1 flu cases in the United States emerged in the U.S.-Mexico borderlands, in San Diego, California, and San Antonio, Texas. Teenagers on spring break in Mexico returned from their revelries with the virus in their bodies. Unable to trace all moments of possible contagion, their schools and universities closed. Around the same time Japan and China instituted quarantines, targeting symptomatic travelers returning from Mexico. The presence of the virus in Mexico characterized the biological landscape of the nation, creating stigma for the country through the presence—and lack of management—of microbes. Eventually, the World Health Organization set the alert for pandemics to its highest level, meaning that the H1N1 virus could be found in most countries and was capable of passing from human to human.[33] The United States declared a health emergency to make stockpiles of antiviral medications, Tamiflu and Relenza, available for distribution. Although travel to and from Mexico was never restricted, the government did employ passive surveillance on the border, stopping people who appeared to be ill and questioning them about their biological status.

Each nation weighed the economic cost of slowing circulations of people and goods against the potential losses from widespread disease. An April 27, 2009, article in *Newsweek* predicted a low death rate from the disease but high potential for economic loss, citing "the fact that the new disease seems not to be terribly virulent outside of Mexico."[34] By this articulation. the political borders of the nation-state defined the virulence of the disease, raising the possibility that the cause of a deadly flu outbreak lay in the ways Mexico managed its environmental landscapes. Microbiologists offered a different geographic story for the virus when they analyzed its genetic code and divided it into five segments that seemed to have come together as the virus spread around the globe and adapted to new hosts. H1N1 is part North American swine influenza, part North American avian influenza, two parts Asian-European swine influenza, and one part human influenza.[35] The virus is a multinational hybrid adapted to dwell in the bodies of many species. The last mutation might have happened in Mexico or somewhere else on the globe, but with it, the virus developed the ability to spread through human hosts. The genetic makeup of the virus was a product of new materialities of the modern age with its global history recorded in its genes.

This genetic study of the virus led one group of scientists to suggest that H1N1 might have been modified in a laboratory, where such hybridizations can be accelerated by technology, and then escaped the lab. They argued that human involvement in creating the virus was a simpler explanation for its virulence than natural evolution.[36] WHO responded to the claims with a definitive statement that H1N1 was of natural origins. Both of these origin stories allowed people to access knowledge of the disease: the organism was a global hybrid that also attained virulence through its diverse interspecies interactions. In the United States, these notions of rapidity and mutation aligned with a prominent racial politics of immigration and terrorism, naturalizing debates over human rights. One political cartoonist depicted a man carrying a grocery bag labeled "Terror Mart" lifting a fence to allow a green glob of flu virus to ooze over the U.S.-Mexico border. The blob says, "Thanks! I'm coming to America to cause mayhem," to which the terrorist replies, "Funny. Me, too."[37] Another cartoon showed a plump piglet labeled "flu" sailing over a barbed-wired border fence, about to land in a field of cactus on the other side.[38] Whether this pig hopped or was tossed, both cartoons argue that the fences on the border cannot secure the nation against the spread of disease.

In many ways, the H1N1 virus was created for the twenty-first century, perhaps even for the year 2009. The virus's genetic code, sequenced and exposed using modern technology, reached beyond terrorism to speak to fears of globalization and the intermingling of races and cultures. Part pig flu, part bird flu, part human flu, the virus crossed species barriers, jumped national boundaries, and embodied cultural concerns about hybridity and the genetic modification of organisms. The virus's genetic code told a story of transspecies movement that could be read as a progression toward infecting humans, and not just very old, very young, or immunocompromised humans. H1N1 expressed many of the qualities sought by the scientists who worked to build a weapon from a microbe, and its appearance in the global spotlight reminded nations of

FIGURE 18. Ed Stein, "Swine Flu," 2009. Reprinted with permission.

terrorism risks. The out-of-season manifestation of the virus raised questions of laboratory safety, as media and scientists speculated on whether the virus was an escaped laboratory experiment. Vaccines for H1N1 could be created, but their production took time and, as with all vaccination programs, was subject to the politics of access and distribution. Now H1N1 is treated like any seasonal flu, and when scientists judge that it is likely to resurface, vaccines are offered to individuals who have access to health care facilities and can seek treatment to potentially avoid the economic and health repercussions of a flu infection. Flu prevention and the annual ritual of the flu vaccine rehearse the politics of immunity that integrate ideas of class and privilege with the mechanisms and technologies of care. By declaring the 2009 epidemic a public emergency, the government initiated an emergency response that opened the CDC's stockpile of antiflu medications for distribution, showcasing the federal government's authority over the disease event and its caring practices. Cities like Albuquerque rehearsed their emergency response plans when the H1N1 vaccine was finally available, setting up the distribution centers they had created for the bioterrorism preparedness plan as temporary drive-through vaccination centers.

The response to swine flu echoed public health warnings that have been institutionalized since the first sanitation projects sought to sweep away microbial threats. Though the U.S. government never implemented a quarantine or restricted travel, the CDC spread its message of flu prevention widely. An early article in the *New York Times* reminded people to stay home if they felt sick: "Individuals . . . should cover their noses and mouths when sneezing. Healthy people are advised to avoid those who are sick, wash their hands often and thoroughly, and try to stay in good general health."[39]

Though the use of face masks was deemed ineffective, the surgical mask quickly became the cultural marker of the swine flu pandemic. Vendors on street corners peddled masks, and public officials appeared on television modeling facial coverings as a sign of good citizenship. The masks reminded people that life was not normal, that the world was in a state of emergency. Images of couples kissing through masks or mask-covered crowds gathering in churches and airports were featured regularly in media coverage of the swine flu outbreak, promoting the mask as a tool that allowed community rituals while protecting individual life.

Although masks block viral particles from entering the body, they also block people from communicating through facial expression, distancing them from interpersonal engagement. In addition, because wearing a face mask when ill is a common cultural practice in Asian countries, the visibility of masks in the United States further associated the flu with foreign nations and sustained a message of protecting the individual against diseases coming in from abroad. The mask marks the human body as a body at risk, placing individual protection from the environment above all other social needs.

The appearance of surgical masks in communities during the H1N1 outbreak demonstrated the results of a decade of bioterrorism preparedness practice in the United States. Though it did not originate as a bioterrorist attack, the pandemic provoked citizens to act. All levels of government and individual citizens had the opportunity to demonstrate their preparedness efforts. Whether or not they were effective in preventing the spread of flu, the masks demonstrated that citizens were ready to respond to biological threats and take action to preserve their own lives. An infrastructure to manage the biological threats created in the modern age had been mobilized. The work of governance in the modern era manufactured a paradigm for the new natures and produced biological citizens who would sustain these new systems of security and care in a world of unspecified microbial risk. The germ wars were under way.

Conclusion

"Freaked Out Yet?"

An editorial published in 2014 during the global Ebola virus panic addressed the possibility that Ebola might be used as a bioterrorist weapon. The headline questioned readers: "Freaked out yet?"[1] Persistent, self-reinforcing messages of fear cohere around bioterrorism. The activities described in this book have built institutions that acknowledge and respond to cultural fears of nature. Preparedness efforts over the past decade have not diminished the fear of a biological attack or the perception that human society is at risk in such an attack. Indeed, some who have read parts of this manuscript have responded with the logics of fear, claiming they are satisfied to know that the government is responding to bioterrorism. National biosecurity seems rational in a world where nature should be feared and where the inability to control nature is perceived to be a persistent threat to human life and society. I argue that by exposing bioterrorism preparedness activities as reactions to culturally produced experiences of fear and risk, we can open up opportunities to build more intentional institutions, thereby creating the world we want to live in rather than responding to a world we fear we live in.

This book has presented a long history of the cultural fear of microbes, from the fears cultivated by outbreaks of diseases like smallpox and spotted fever to the anticipatory fears that viruses like Marburg and Ebola will escape their containment in high-security laboratories. The human experience of disease produces a sense of victimhood in microbial encounters. A public health poster created for international

FIGURE 19. Poster created by Webber Training.

training on infection control and prevention shows a tiger hunched in the grass, looking off the page with fiery orange eyes. The caption reads, "Predator—Germs; Prey—You. Clean everything . . . disinfect where appropriate." Not only does the poster name microbes as predators, but it associates their encounters with humans with tigers' attacks on their prey. By aligning microbes with large, toothy carnivores, the poster accesses broader fears about human vulnerability. The poster also demands action from its viewer. The same cleaning and disinfecting required of citizens at the beginning of the twentieth century to fight disease in their homes are encouraged today, and public health institutions continue to educate citizens about microbial threats. Adherence to their instructions can be motivated by the notion that germs lurk on every surface like tigers in the grass, waiting to attack humans to serve their own ends.

Bioterrorism accesses this fear of predation in two ways, merging the intentionality of the microbe with the purposes of terrorists and thereby creating cultural understandings of bioterrorism that are naturalized by fears of microbes and politically constituted through a war against terror. Despite more than fifteen years of work in the United States to deter terrorism, the perceived risk of a bioterror attack has not abated. The unknowability of the future produces a perpetual state of risk in the present, which can be addressed only by somehow making the future knowable. Scenario planning exemplifies how scientists and policy makers work to make the future known in order to intervene in the present. Others rely on expert opinions to judge the likelihood of biological attack. A study published in 2015 assessing the probability of a biological attack surveyed experts who "held responsibility for shaping public policy at the nexus of life science and national security or on the basis of their expertise."[2] Only one in fifty-nine saw no likelihood that biological weapons would be used in the next ten years, and twenty-nine respondents suggested at least a 75 percent likelihood, on a scale from 0 to 100 percent, that there would be an attack causing at least one hundred people to fall ill. More than forty of the experts proposed at least a 50 percent likelihood of biological attack, most likely by a nonstate entity (particularly religious extremists) or a covert attack by another state.

Given the motivations experts have in sustaining the perceived risk of attack, and the reluctance of scientists ever to describe risk as zero, this accounting of a moderate risk of bioterrorist attack is not unexpected. The survey raises the question, however, of whether the government response to the threat of bioterrorism matches the expert calculation of the risk of attack. The calculations of risk may never be clear enough to justify the expense of preparedness; these actions are sustained by deeper ideologies that generate compliance among the governed. Though scientists reason that bioscience research on deadly microbes has multiple uses, and public health agencies overhaul their infrastructure for the sake of general improvements to citizens' health, these activities have the direct purpose of preventing and responding to bioterrorist attacks. Because this work is rooted in deep fears and desires to control nature, it has the potential to continue unchecked for generations.

MICROBE MATTERS

The technoscientific and political practices of the last century have made microbes in ways that produce fear and shape the interactions that humans have with each other and with the world around them. These politics of nature permeate modern life, as people build infrastructure and design social systems to manage microbes. Bioterrorism preparedness practices exemplify how cultural fears of nature sustain governments that manage people and the environment to contain biological risk. Modern citizens are biological citizens whose hybrid interspecies bodies contain more microbial DNA than human DNA and whose bare life is sustained by microbes.[3] Thus the making of the microbe has simultaneously remade the human body in terms of a relationship in which microbes both constitute and threaten human life.

The history of bioterrorism presented in this book shows how new fears coalesced around microbes through the work of scientists and citizens, who in turn created new mechanisms for managing microbial nature. Microbes have been made through the belief that nature can be managed according to human desires. Biological laboratories, vaccination programs, and government agencies all manifest our social conviction that human life can be preserved by controlling nature. The mechanisms societies have developed to manage macrobiological nature were also formed in part to contain and control microorganisms, but those who study nature have largely ignored the management of microbes. Including microbes in the critical study of nature elucidates how institutions of war, science, and health are built around ideas of nature.

People rely on science to expose the materiality of microbes and to explain how to manage our relationship with them. Scientific practices have made microbes into enemies. We know germs as minuscule organisms that can be battled with bleach, antibiotics, and vaccines, and also as life forms that can be managed by humans to wage war. Scientists have shown us how microbes can be deliberately dispersed through air or water, moved through the population, and altered in a laboratory to target vulnerabilities in human bodies or social systems. Weaponization has changed the geography of microbes through mass production and stockpiling, as well as the through creation of laboratory spaces that cultivate and contain organisms that hold the potential to be used as weapons. Bioterrorism subverts the geographical knowledge of microbes that associates emerging infectious disease (a code for diseases that scientists have not studied and made knowable, or that have transcended an evolutionary boundary to emerge as human threats) with equatorial regions and the tropics, breaking down natural boundaries that contain disease and locating the birthplace of deadly pathogens in high- and low-security laboratories, not just in jungles and animal bodies. By bringing this newly materialized microbe into acts of nation building, bioterrorism naturalizes actions of individuals and governments, creating a twenty-first-century biopolitics centered on fear and an unending search for security.

Microbes have also changed what it means to be human and a member of a nation. We have created institutions to manage microbes, ranging from biosafety laboratories to government agencies, and social practices to contain and control microbes, such as substituting fist bumps for handshakes during flu season and applying hand sanitizer when we enter a building. Our political systems were built in part to contain microbes, and these efforts in turn define the terms of good citizenship. Ideas about security, community, and nationhood centered on the microbe are changing people's lives.

Science practices also make explicit the outcomes of this political work, as scientists and policy makers collaborate to calculate biological risk, imagine social futures after biological events, and create systems that promise to secure populations against microbial threats. Simulation exercises bring scientific practices into the work of policy making, predicting how human interactions might move microbes. Government agencies have expanded to produce more knowledge about microbes and new ideas about how to manage microbial risk by managing citizens' behaviors. Indeed, nations define boundaries in terms of biological risk and manage the movement of bodies across borders as an act of national security. Humans' fear of microbes has motivated massive government expenditures and rationalized extraordinary political interventions in social lives. These outcomes of the effort to manage microbes offer a vital, but often overlooked, site for studying the cultural politics of nature.

The natural history of the microbe as constituted through bioterrorism preparedness gives evidence of broader shifts in ideologies about nature. The war on terror has been constituted not only through a reworking of the ideologies and mechanisms of war but also by conditioning citizens to live in a world of unspecifiable risk. Biological terrorism entwines cultural fears of nature with the politics of war; it engages the institutions of science, particularly the life sciences, in the work of nation building and national security. This work, in turn, brings the economics, knowledge production, and social mitigation practices of disease control into the national project to secure citizens' lives.

I do not seek to discredit government efforts to decrease human suffering caused by disease or to create healthy and resilient societies. However, the expenditure of resources in one area may create deficits in others, and the effects of this work are not balanced. The social effect of bioterrorism is to cast deliberate political decisions as inevitable responses to a world full of natural and living risks. By showing how bioterrorism is culturally constituted and scrutinizing the institutions that make biological terrorism knowable, I aim to recognize how systems of governance are rooted in longstanding ideologies of nature and to open the possibility of new natures and, consequently, new futures.

Twentieth-century practices in public health and international warfare made germs for the modern age, along with the belief that microbes could be managed according to human desires. The alliance of war and science technologized microbes both by creating weapons and by creating biosafety laboratories where microbes can be contained. Biological warfare remakes vulnerability in the face of natures that have been altered to be more virulent, contagious, harmful, and far-reaching. By the time the United States disbanded its offensive bioweapons research program in the 1970s, the decades-long alliance between the science of war and the study of life had vastly expanded popular imaginings of how microbes might harm people.

These imaginings stirred after September 11, 2001, when envelopes stuffed with anthrax spores passed through the federal mail system. In response, the United States enlisted its government agencies and the life science industry in a war on terror. The effort to secure the nation remade these social and government institutions. Funding for bioterrorism preparedness changed the interactions between government and citizens: tools previously used for public health surveillance and scientific research became weapons in a war against terrorism. In consequence, these industries instituted new ways of thinking about human lives and calculating risks in the system of exchanges between humans and microbes. The knowledge-producing mechanisms of science responded to the need to calculate the risk posed by the presence of microbes in the landscape. Imagining the effects of bioterrorism through systemic study creates new forms of scientific knowledge and new future knowledge of nature. In an era of security, it is particularly important to consider how predictions of the future are used to rationalize present behaviors that manage human bodies, individually and collectively.

Bioterrorism affirms and undermines boundaries. Microbes defy boundaries between the human and nonhuman as they move in and out of our bodies. Bioweapons science straddles the line between offense and defense, potentially eroding the ethical boundaries of warfare.[4] And the boundaries of the modern nation-state have biological dimensions that must be policed. National biosecurity is a set of practices that regulate the flow of goods, services, and human and nonhuman life. Nations work diligently to secure their borders, which are, in reality, crossed repeatedly by all forms of life, including millions of unseen microbes. The policing of life becomes an act of care for the nation. Intimate and invasive acts are justified as vital to public health and national security, even as they degrade individuals and naturalize racial discourses about cleanliness and contamination. The management of biological nature becomes an act of nation building, and containing microbes becomes a vital act of national security.

If we posit an endless war against an invisible threat, society will continue to expend time, energy, and money on the germ wars. Bioterrorism has brought the life sciences into the service of the war industry and biosecurity into the work of nation building. This alliance between war and biology has reset the terms of terror in the genomic age, creating a set of cultural fears that are naturalized in the form of the microbe. This new landscape of fear provides a backdrop for innumerable acts of governance aimed at securing human life. The twenty-first century germ wars are waged in laboratories, public health clinics, border crossings, airports, government offices, and media institutions. Though we imagine we are battling invisible, highly mutable microbes, the germ wars might best be understood as the struggle to delineate human society in a world fully occupied by microbial life. Like any war, the enduring outcome will be the society we build around these new forms of nature.

Acknowledgments

During my decade-long career with the federal government, colleagues often entreated me to keep my work life separate from my personal life, an admonition that seemed impossible when I was snowed in with a dozen coworkers for a winter in Yellowstone Park, fifty-five miles by snowmobile from the nearest town. It has been equally impossible during the many seasons of writing this book. I have marveled at the porous boundaries between the spaces for researching, writing, and working and those for nurturing, talking, and playing.

This book was written in five states and researched in five more. Over and over, I carried the scrappy trappings of an ethnographer to a new community and created a home among new neighbors, mentors, and colleagues. In every state I found that my own sense of home was being reworked and refined every bit as much as my research project. Because my debts to the people named below are deeply personal as well as professional, I acutely sense my inability to recompense them for the complete life they have offered me. I hope they will recognize this book as my effort to make something worthy of the care they have given this project and that they will forgive its shortcomings.

My early mentors in New Mexico created a community that provoked, challenged, and motivated scholarship while cultivating humanity and collegiality. Jake Kosek helped me find paths to answers that mattered. His vigor, vision, and generosity sustained me from near and far, and I am profoundly grateful for his belief in this project and in the worthwhile pursuit of critical scholarship. Alyosha Goldstein smilingly and patiently offered the pep talks that are vital to completing a research project, along with ready insights that took this work to important confluences. Rebecca Schreiber guided me into a new discipline and shaped my academic ways of seeing. David Correia infused this project with energy and perspective, and I thank him for his willingness to join in this venture.

My appreciation extends to many other scholars who have mentored my academic career, including Amanda Cobb, Virginia Scharff, Norma Pecora, and the late Timothy Moy. Jennifer Richter and Carson Metzger scrutinized, tested, poked, and prodded my ideas from beginning to end. Kristan Cockerill has been a teacher turned collaborator, quick to scaffold my musings about science and society. Char Peery and Kari Schleher enriched this work with keen cross-disciplinary perspectives. The 2003 American Studies Cohort created safe spaces for inquiry and brought joy to graduate school. I add to this list a group of New Mexicans whose character, intelligence, and good humor has enlivened my writing and my life in countless ways: Andrew Bateman, Lynne Charapata, Buckner Creel, Michelle Croasdell, Bill Dewan, Alison Fields, James Grubel, Chuck Hayes, Donna Knaff, Kathryn Lenberg, Craig McClain, Jordan Okie, Sandy Rodrigue, Christopher Shank, and Jason Thomas.

My colleagues in California, at UC Davis and UC Berkeley, challenged me to situate my work in history and geography. My special thanks to Julie Sze, Michael Ziser, Louis Warren, Traci Brynne Voyles, and the many participants and attendees of the Environments and Societies Colloquium. The year we spent sitting around a table, reading, talking, and feasting, helped me wrestle with my own work in progress. A year in Utah brimmed with collaboration. My thanks to Christy Clay, Brent Olson, and Tiffany Rivera for the chance to learn in the classroom (or in a fifteen-passenger van), and to Hikmet Sidney Loe and Connie Kitchens for creating an interdisciplinary collaboration in the hallway for disciplinary orphans. As I put the finishing touches on the manuscript in Colorado, I am building yet another academic home, and I am grateful every time I set foot in Kelley Hall for my colleagues in the Master in Environmental Management program at Western State Colorado University.

For years I envisioned my seasons working in national parks as the perfect retreats for writing a manuscript. While it sometimes felt like the hiking, camping, and musings around the campfire distracted from my scholarly pursuits, those interactions resonate throughout this book. My thanks to colleagues in Yellowstone and Canyonlands for believing in this peculiar bioterrorism project as much as they supported my career in the "green and gray." I particularly thank Sharon Brussell, Janis Buckreus, Sierra Coon, Jack Foy, Laura Goforth, Denise Herman, Eric Johnson, Justin Monetti, Trudy Patton, Erika Reinicke, Jamie Silberberger, Michael Stuckey, Ted Wilbur, Mary Wilson (with additional thanks from my mother), Gretchen Wise, and Will Yaworsky, along with Raquel Romero and members of the 2013 GOAL Academy.

Though I logged time at the keyboard in these five states, books are patched together from ideas carried around the world to conferences, symposia, and meetings. I have presented this work on panels in disciplines ranging from American Studies to microbiology, and I am indebted to countless critiques from fellow panelists and audience members. I am particularly grateful for the holistic learning I experienced with colleagues at the Southern California STS Retreat; the Workshop for the History of Environment, Agriculture, Technology, and Science; and the Clinton Institute for American Studies Summer School. The women of Mapping Meaning have centered me as an academic, park ranger, and human being. On our weeklong trip down the Green River, I tapped out part of this book on a tablet powered by a solar panel, and we laughed as we cast giant shadows on a canyon wall. Truly, this is a book of many places.

More than anything, though, this is a book *about* places, and I thank the many scientists, citizens, policy makers, and participants who introduced me to their places and gave their time and expertise to the project. It was a privilege to sit down with so many knowledgeable and interesting people, and their stories are the life of this research. I thank numerous individuals, named and unnamed in the text, from the Centers for Disease Control and Prevention, Rocky Mountain Laboratories, the Albuquerque Office of Emergency Management, the Southwest Border Food Safety and Defense Center, and the Energetic Materials Research and Testing Center. I thank the archivists and librarians who joined me in an archaeological investigation of government and science, casting new light on old ideas or reviving old ideas with new life. My particular thanks to the Center for Southwest Research, the RML library, and Jeff Karr with the Center for the History of Microbiology/American Society for Microbiology Archives. Finally, I thank the people of Georgia, Maryland, Montana, New Mexico, and California who gave time to articulate their wide-ranging ideas about how bioterrorism had shaped their lives. I hope I have told their stories well.

Special thanks to the editorial team at UC Press for their patience and care. I am awed and humbled by the thoughtful, detailed critiques of the manuscript given by Kathleen Vogel, Traci Brynne Voyles, and two anonymous reviewers. I thank Erika Büky for revitalizing my prose with remarkably wise and rigorous editing.

The research and writing of this book was supported by a Mellon Fellowship with the UC Davis research initiative Environments and Societies: History, Literature, and Justice, as well funding from the National Science Foundation. A grant from the American Society for Microbiology supported research in Maryland. At the University of New Mexico, the Graduate and Professional Student Association, Office of Graduate Studies, Department of American Studies, and a Regent's Fellowship funded parts of this research and its circulation at academic conferences.The breadth of this work is due to the opportunities their funding provided.

Finally, there are many who have allowed me to bring my work into their lives, shattering any boundaries they may have wished to preserve around their own time and labor. These friends and family sustained me through my many moves, often laboring physically to carry crates of books and files as I set off for the next location. My family is tremendous in its size and heart, and they teach me daily about generous living. My parents, Bill and Sheila Armstrong, have invested wholly in my learning, from the earliest days of carpooling, science fairs, and family vacations. My reliance on their wisdom has only increased with time. My grandmother, Elna Lucas, read even the roughest of manuscripts. William and Camille Armstrong offered housing, meals, and cheerful chauffeur service to and from the archives. Karen Henker Garthwait buoyed me through fateful years of wandering—geographically and ideologically—providing a sofa and a solid place to land anytime I needed one. Known to be quick with a comma, Scott Clark also brandished hidden culinary skills in an hour of need. I am immensely grateful that he endured with me.

When I first sat in a seminar room with Shannon McCoy-Hayes, Jennifer Richter, and Carolyn McSherry, I could not have imagined how entangled our lives would become. Real life is sitting with you atop a sand dune, eating Nutella, and talking about biopolitics. To Jen, I add my particular thanks for guiding me through my many moments of terror and insecurity. This book, and so many good things, would not exist without you.

Notes

INTRODUCTION

1. Williams 1980.

2. Bennett 2010.

3. "UPI Poll: Bioterrorism Seen as Top Threat," United Press International, www.upi.com/Top_News/2007/02/23/UPI-Poll-Bioterrorism-seen-as-top-threat/UPI-51211172268000/?st_rec=5271957067200, accessed February 25, 2007.

4. The CDC estimates the cumulative number of deaths of people diagnosed with AIDS in the United States was 617,025 in 2008, with about 16,000 deaths annually. "HIV/AIDS: Statistics and Surveillance," CDC, accessed January 23, 2010, www.cdc.gov/hiv/topics/surveillance/basic.htm#ddaids. For a history of biological attacks, see Barras and Greub 2014.

5. Calculating annual government expenditures for biosecurity is difficult. Calculating line items marked explicitly for bioterrorism ignores funds allocated to vaccine research and public health improvements. The numbers cited here borrow from Jeanne Guillemin (2005), whose work explicitly tallies expenses. Allocations by fiscal year can be accessed at *Budget of the United States Government*, www.gpo.gov/fdsys/browse/collectionGPO.action?collectionCode=BUDGET.

6. Simon 2001.

7. Rachel Nowak, "Killer Virus," *New Scientist*, January 10, 2001. An early example of a biological attack, often recounted in American military lore, tells of British soldiers passing smallpox-infected blankets to Native Americans in order to "extirpate this excrable [*sic*] race" (Nester 2000, 115.) For a cultural history of smallpox, see Preston 2000; Koplow 2003; Tucker 2002; Peters 2005. This topic is explored further in chapter 1.

8. Preston 1994; Garrett 1994.

9. Miller, Engelberg, and Broad 2001.

10. Masco 2008, 362.

11. Massumi 2007, 9.

12. Lakoff 2007.

13. Coole and Frost 2010, 19.

14. Louis Pasteur, "Germ Theory and Its Applications to Medicine and Surgery, 1878," in *Scientific Papers: Physiology, Medicine, Surgery, Geology, with Introductions, Notes and Illustrations*, trans. H.C. Ernst (New York: P.F. Collier & Son, 1910), 359–60.

15. The process of heating food to a temperature that kills microorganisms, *pasteurization*, bears the name of the scientist who showed how to purify through the eradication of microbial life.

16. Thanks to Christopher C. Shank for this insight. See also Blank 1999.

17. Margulis 1998, 111.

18. Jane Bennett argues that we insist on an ontological divide between people and objects because we fear losing the moral grounds for privileging humans over germs, for example. Bennett 2010, 47.

19. Rozsa 2000.

20. Masco 2011.

21. *Department of Defense Fiscal Year 2015 Science and Technology Programs: Pursuing Technology Superiority in a Changing Security Environment, Hearing Before the House Committee on Armed Services*, 113th Cong. (2014) (statement of Arati Prabhakar, director, Defense Advanced Research Projects Agency). Posted on DARPA's Facebook page, www.facebook.com/DARPA/posts/10152419782972150, accessed 7 August, 2016.

22. National Research Council, *Biotechnology Research in an Age of Terrorism: Confronting the Dual Use Dilemma* (Washington, DC: National Academies Press, 2003), 19.

23. Centers for Disease Control and Prevention, "Bioterrorism Overview," http://emergency.cdc.gov/bioterrorism/overview.asp, accessed 22 August 2016.

24. *Strategic Perspectives on the Bioterrorism Threat: Hearing Before the House Emergency Preparedness, Response, and Communications Subcommittee, Infrastructure Protection*, 114th Cong. (2015) (statement by Jim Talent, senior fellow, Commission on the Prevention of Weapons of Mass Destruction Proliferation and Terrorism). In the words of the virologist Mark Collett, "Nature is a current and ongoing source of these viral pathogens" and "has provided would-be bioterrorists an ample supply and selection of quite virulent viruses." Collett 2006, 86, 94.

25. By this conception, life is the outcome of the interactions of species, particularly the interaction of labor. Karl Marx wrote that labor "is a necessary condition . . . for the existence of the human race; it is an eternal nature-imposed necessity, without which there can be no material exchanges between man and Nature, and therefore no life" (Marx 1973, 26). The motivation for humans to labor lies in the presumption that without labor, life would cease. Marx suggests that labor also creates the possibility for humans to work for the survival of the group. Marx and Engels (1975, 584) state broadly that at the moment labor begins, humans distinguish themselves from animals, and without human production of nature, life would not exist: "The most that the animal can achieve is to *collect;* man *produces*, he prepares the means of life, in the widest

sense of the words, which without him nature would not have produced. This makes impossible any unqualified transference of the laws of life in animal societies to human society." As a result, the production of life has been a politics of estrangement, creating a social system that values human life above others in order to convince humans that their labor is vital. The Marxist view of nature insists that the laws that govern humans are not the same as the laws of biology, chemistry, or physics. Human life cannot be explained by biology.

26. Wills 1998.

27. Schell 2002.

28. Raney 2003, 392.

29. See for example Brandt 1987; Tomes 1999; Kraut 1994; Tomes 2002.

30. Lewontin 1991, 16.

31. Prigogine 1999.

32. Quoted in Margulis 1998, 23.

33. Safina 1997, 142.

34. Margulis 1998, 79.

35. Braun and Whatmore 2010, xxi.

36. Margulis 1998.

37. Ibid., 7.

38. Bennett 2010, xix.

39. Cooper 2008, 3.

40. See for example, Rajan 2006; Fortun 2001; Latour 1999.

41. Braun 2007, 7.

42. Petryna 2002, 10.

43. Gusterson 2008, 551.

44. Robin 2004.

45. Bourke 2006, 13.

46. The belief that humans are agents of harm characterizes the concept of risk. Niklas Luhmann (1993) contrasts risk with danger, claiming that the latter expresses harm from an external force (such as a natural disaster), whereas risk comes about because of people's actions. In response to Luhmann and Hacking (2003), Collier, Lakoff, and Rabinow (2004) ask whether human responsibility for mitigating harm is also inherent to the definition of risk.

47. Beck 1992; Giddens 1999.

48. Adriana Petryna (2002, 7) uses this phrase to describe how the Soviet government shifted responsibility for seeking reparation for harm to the citizens who had been affected by the nuclear meltdown at Chernobyl.

49. Nicholas King (2004) considers this issue in his study of the emerging-diseases public health campaign, arguing that the campaign turned social ambivalence about modernity into consensus on new risks and necessary interventions.

50. Lakoff 2007.

51. Braun 2011, 390, 403.

52. Examples of ethnographic studies that address questions of biopolitics include Fortun 2001; Petryna 2002; Rabinow and Dan-Cohen 2005; Tsing 2005; Rajan 2006.

53. Petryna 2002, 220.

54. Rajan 2006.
55. Bush and Perez 2012, 42.
56. Rozsa 2014.
57. Ibid., 129, 131.
58. For studies of the history of biological weapons, see, for example, Miller, Engelberg, and Broad 2001; Wheelis, Rozsa, and Dando 2006; Barras and Greub 2014.
59. Rozsa 2000.

CHAPTER 1. "SMALLPOX IS DEAD"

1. "Strengthening the International Regime against Biological Weapons," White House press release, November 1, 2001, https://georgewbush-whitehouse.archives.gov/news/releases/2001/11/20011101, accessed August 8, 2016.
2. "President Increases Funding for Bioterrorism by 319 Percent," (speech by George W. Bush, University of Pittsburgh, Pittsburgh, PA, February 5, 2002), http://georgewbush-whitehouse.archives.gov/news/releases/2002/02/20020205-4.html, accessed August 8, 2016.
3. Ibid.
4. Jasanoff 2010, 15.
5. For a cultural history of smallpox and detailed biological information, see Koplow 2003; Peters 2005; Tucker 2002; Preston 2000; Fenn 2001; Henderson 2009; Hopkins 1983.
6. Stern 2002, 102.
7. Crosby 1986.
8. For a broader critique, see Correia 2013; Sluyter 2003; Robbins 2003.
9. See, most famously, Jared Diamond, *Guns, Germs and Steel: The Fates of Human Societies* (New York: Norton, 1999).
10. Hitchcock 1909, 395.
11. Thanks to Traci Brynne Voyles for this insight.
12. Crosby 1986; see also McNeill 1976.
13. Peckham 1947, 226; Anderson 2000, 809 n; Ranlet 2000, 427–28.
14. In the narrative, Bouquet suggests that blankets might spread the disease and replies to Amherst accordingly. Amherst responds, "You will do well to inoculate the Indians by means of blankets, as well as every other method that can serve to extirpate this execrable race." Anderson 2000, 809 n.; Grenier 2005, 144; Nester 2000, 114–15.
15. Quoted in Dixon 2005, 151.
16. Behbehani 1983.
17. Jenner 1798.
18. In 1881, when Pasteur developed the Anthrax vaccine, he kept the name *vaccination* to refer to all processes of inoculation against disease. Behbehani 1983, 470.
19. Razzell 1977.
20. Behbehani 1983; Behbehani 1980.
21. The governor of the Council of the Indies noted that depopulation would mean a decrease in income from taxes that the Spanish received from the colo-

nies as well as a decrease in commerce and farming. The economic benefits of a vaccination mission would, in his view, justify paying for it from the royal treasury. Three times during the expedition, Balmis's group docked in a town only to discover that vaccination had already been introduced, testifying to the rapid global spread of the technique. Hopkins 1983; Smith 1974; Bowers 1981.

22. John G. Cotton, ed., *The Medical Intelligencer: Devoted to the Cause of Physical Education and to the Means of Preventing and of Curing Disease*, vol. 4 (Boston: John Cotton, 1827), 72. Smith earned no salary, though he received free postage for the distribution of the vaccine. Although he was authorized to charge a fee—his usual price was five dollars—he struggled with the ethics of withholding the vaccine from those who were unable to pay.

23. "Seventeenth Congress—First Session: In the Senate. House of Representatives. Thursday's Proceedings—March 28," *Niles' Weekly Register (1814–1837)*, March 30, 1822, 80.

24. "Seventeenth Congress—First Session, House of Representatives" *Niles' Weekly Register (1814–1837)*, April 20, 1822, 125. The committee report argued that because the task assigned to the vaccine agency was so vast, it could not be done by the agent without producing a decline in the quality of the product.

25. Untitled article, *Niles' Weekly Register (1814–1837)*, March 2, 1822, 1; Untitled article, *Niles Weekly Register (1812–1837)*, April 6, 1822, 81.

26. E. Cohen 2009, 63.

27. Ed Cohen argues that "only with the advent of biological immunity does a monadic modern body fully achieve its scientific and defensive apotheosis." Ibid., 8.

28. Events recorded in the committee's official report to WHO on the smallpox eradication program. Fenner 1988.

29. The 1967 resolution also dedicated about 5 percent of the WHO budget, roughly $2.4 million annually, to the program and established a headquarters for the directive.

30. Lakoff 2012.

31. Fenner 1988, 421.

32. Henderson himself denied military tactics: "Many persons inside and outside WHO mistakenly concluded that the achievement [of global eradication] could be attributed to a generously financed, enthusiastically supported and authoritatively directed programme similar to a military campaign. That the programme had none of the characteristics is apparent." Ibid.

33. Jitendra Tuli, "India's 'War Plan,'" *World Health: The Magazine of the World Health Organization*, May 1980, 13.

34. Ibid.

35. Marcella Davies, "A Job Well Done," *World Health: The Magazine of the World Health Organization*, May 1980, 6–10.

36. Tuli, "India's 'War Plan,'" 13.

37. Davies, "A Job Well Done," 7; Donald A. Henderson, "A Victory for All Mankind," *World Health: The Magazine of the World Health Organization*, May 1980, 4.

38. Tuli, "India's 'War Plan.'"

39. Henderson, "A Victory All Mankind."

40. Tuli, "India's 'War Plan.'"

41. E. Cohen 2009, 63.

42. Arun M. Chacko, "A Goddess Defied," *World Health: The Magazine of the World Health Organization,* May 1980, 15.

43. Foucault 2007, 59.

44. The vaccine is effective only for a decade, though the scar lasts much longer, and only about 80 percent of smallpox cases leave facial scars, so this surveillance could not be considered foolproof.

45. This method was considered by many to be an inefficient means of vaccination and a waste of the officials' time, for many more bodies could be treated if they came to a central location for vaccination.

46. Foucault 2007, 66.

47. World Health Assembly Resolution, *Declaration of Global Eradication of Smallpox,* WHA 33.3, 1980, 1, May 8, 1980, http://apps.who.int/iris/bitstream/10665/155528/1/WHA33_R3_eng.pdf.

48. Henderson, "A Victory for All Mankind," 5.

49. Davies, "A Job Well Done," 8.

50. Yugoslavia declared martial law in 1972, when an outbreak occurred in Kosovo Province. The state enacted a nationwide vaccination plan and held exposed individuals in quarantine.

51. Koplow 2003, 187. Scholars and practitioners have theorized numerous reasons why a second eradication event has not occurred. Perhaps the horror of smallpox compelled action in ways that other diseases do not. Other explanations might be cultural. Marcos Cueto (2007), for example, argues that people in Mexico do not see malaria as a problem, or at least not one attributable to a single cause, and are more concerned with other health issues. Despite repeat bursts of funding for eradication efforts, malaria has not been eliminated.

52. World Health Assembly Resolution, *Smallpox Eradication: Destruction of Variola Virus Stocks,* WHA 52.10, 1999, 1–4, May 24, 1999, http://apps.who.int/iris/bitstream/10665/79354/1/e10.pdf.

53. Current WHO-sanctioned research focuses on antiviral agents and vaccines as well as detection mechanisms. A 1999 resolution outlined six areas of research being conducted with the virus stockpiles, including "phylogenetic analysis . . . nucleotide sequence analysis of variola virus DNA, serological detection of variola virus, antiviral agents and animal models of smallpox." World Health Organization, *Advisory Committee on Variola Research,* Report of the *Second Meeting,* 15–16 February 2001.

54. Laboratory research, however, still carries risk. The last person to die of smallpox was a scientist in England who contracted the disease in a laboratory accident in 1978. Twice a year, WHO inspectors tour the reserve laboratories to assess the security of the facilities and the safety practices of workers. When WHO again voted to delay the destruction of the virus in 2007, claiming that "the destruction of all variola virus stocks is an irrevocable event and that the decision of when to do so must be made with great care," a committee halfheartedly reminded WHO delegates that previous assemblies had voted to destroy the reserves, urging them to complete the task their predecessors had begun in the

1950s: the obliteration of living smallpox virus. World Health Organization, *Executive Board, 120th session, Resolutions and Decisions Annexes*, January 22–29, 2007, 46–8. http://apps.who.int/gb/ebwha/pdf_files/EB119–120-REC1/p3-en.pdf, accessed August 8, 2016.

55. World Health Assembly resolution, *Smallpox Eradication: Destruction of Variola Virus Stocks*, WHA 60.1, 2007, 1–3, May 18, 2007, http://apps.who.int/iris/bitstream/10665/22569/1/A60_R1-en.pdf.

56. The Federation of American Scientists' "Fact Sheet on Smallpox," for example, sustains the rumor of countries' harboring smallpox reserves. Federation of American Scientists, *Smallpox Fact Sheet, Variola major, V. minor*, www.fas.org/programs/bio/factsheets/smallpox.html, accessed August 10, 2016. See also Preston 2000, 24; Tucker 2002, 209–19.

57. Cello, Paul, and Wimmer 2002.

58. James Randerson, "Did Anyone Order Smallpox?" *Guardian*, June 23, 2006.

59. Rachel Nowak, "Killer Virus," *New Scientist*, January 10, 2001.

60. Ibid. See also Jon Cohen, "Designer Bugs," *Atlantic Monthly*, July–August 2002, 113–21; "Killer Virus Accidentally Made in Lab," *Current Science* 86, no. 14 (2001); Miller, Engelberg, and Broad 2001.

61. J. Cohen, "Designer Bugs."

62. The center is now known as the Invasive Animals Cooperative Research Centre. Jackson had theorized that as the mouse's immune system fought off the virus, it would produce antibodies that would also attack the mouse's eggs, causing sterility in the female. The sterile mice could still mate but over time would dilute the effective breeding of the mouse population and thereby offer some relief from the mouse plagues that overrun the nation about every four years.

63. Humans cannot contract mouse pox, though the strain is similar to the *Variola* viruses that cause smallpox. However, because of the widespread use of mice in science research, an outbreak of mouse pox could undermine years of work by killing millions of laboratory mice. Thus, the United States prohibits experiments with mouse pox, and Jackson and Ramshaw were working in one of a few facilities in Australia that allows work with the virus.

64. J. Cohen, "Designer Bugs."

65. Ibid. Ramshaw consulted with Frank Fenner, who was his colleague at the John Curtin School of Medicine and the coauthor, with D. A. Henderson, of a definitive history of smallpox, *Smallpox and Its Eradication* (1988). Jackson consulted with the CRC director Bob Seamark. Seamark consulted with the government agency that funded his program, which in turn took the question to the Australian Department of Defence.

66. CSIRO Australia, "Discovery Prompts Call for Biowarfare Review," ScienceDaily, January 15, 2001, www.sciencedaily.com/releases/2001/01/010110180440.htm, accessed August 8, 2016. Similar events in 2011 surrounded a decision to publish research about a strain of avian flu that was modified in a laboratory to survive airborne transmission. Again, the debates centered on the practices of open science and the value of publication to the science community versus the risks of knowledge being used by individuals to inflict harm on populations.

67. Humans are susceptible to four poxviruses: cowpox, monkey pox, vaccinia, and variola. (Chicken pox is misnamed: it is not a true poxvirus, and it infects humans but not chickens.)

68. See Wald 2002b.

69. Colonies also passed laws against the practice because of the risk of contagion and spread of the disease among civilians. Letter from George Washington to Major General Horatio Gates, 5–6 February 1777, in *The Papers of George Washington,* ed. Philander D. Chase, Dorothy Twohig, Frank E. Grizzard, and Edward G. Lengel (Charlottesville: University Press of Virginia, 1985), 8.

70. "Remarks by the President on Smallpox Vaccination," White House, Office of the Press Secretary, December 13, 2002. Statement by President George W. Bush. https://georgewbush-whitehouse.archives.gov/news/releases/2002/12/20021213–7.html, accessed August 8, 2016.

71. Ibid.

72. When the military mandated anthrax vaccinations for all personnel in 1997, many refused the vaccine because of its side effects. Air Force Major Sonnie Bates was honorably discharged when he refused the vaccine in 1999.

73. "Remarks by the President on Smallpox Vaccination."

74. The *New York Times* reported that individuals who wished to receive the vaccine immediately could enroll in clinical trials, a suggestion put forward by Dr. Julie Gerberding, director of the Centers for Disease Control and Prevention in Atlanta. "As this program unfolds, vaccine safety is a top priority," Gerberding said. "We intend to do everything that we can to minimize the risk." Richard W. Stevenson and Sheryl Gay Stolberg, "Threats and Responses: Vaccinations: Bush Lays Out Plan on Smallpox Shots," *New York Times,* December 14, 2002.

75. Lawrence K. Altman and Denise Grady, "Threats and Responses: The Smallpox Vaccine," *New York Times,* December 15, 2002.

76. Kaltman et al. 2006.

77. Masco 2011.

78. Reid 2006. Disciplining bodies is a principal act of war, for the human body is a resource for preparedness. As Ed Cohen argues, such disciplining is also the basis for the naturalization of "the military model as the basis for organismic function." E. Cohen, 2009, 20.

79. "Dr. Ken Alibek's Immune System Support Formula," www.drkenalibek.com, accessed May 12, 2008. This website, www.kenalibek.com, and www.dralibek.com (since lapsed) served as the online sales points for Alibek's pills. The marketing pitch can be found through the archive of the Vital Basics: Advanced Natural Health Formulas page at the Wayback Machine, https://web.archive.org/web/20030409074047/http://www.vitalbasics.com/default.asp, accessed August 10, 2016.

80. In an intriguing metaphor for medical perspectives on immunity, Polly Matzinger describes popular conceptions of the human body as imitative of the political devices of civil defense: "For half a century we have studied . . . models in which immunity is controlled by the adaptive immune system, an army of

lymphocytes patrolling the body for any kind of foreign invader. Recently there has been a shift to include the cells and molecules of the innate immune system, an army of cells and molecules patrolling the body for a subset of foreign invaders that are ancient enemies. . . . Perhaps it is time to stop running a cold war with our environment." Quoted in E. Cohen 2009, 28.

81. The languages of war when applied to the human body's relationship with the world disconnect people from nature. This sustains the condition of modernity described by Latour (1993), in which separating the natural world from the cultural world seems to guarantee that humans can act freely for their own destiny.

82. Ken Alibek. "Immune Support: Understanding the Connection Between Immunity and Health," www.drkenalibek.com, accessed May 12, 2008. See also Martin Enserink, "Biowarrior Branches Out," *Science Now*, October 9, 2002, 4.

83. Critics questioned Alibek's motivations because he kept close ties to the business side of biodefense. Claims he made about the smallpox vaccine's providing protection against HIV were contested on both scientific grounds and because of Alibek's financial interests. In 2006, Alibek resigned from a tenured post at George Mason University, where he founded the nation's first biodefense graduate education program, to become the CEO of AFG Biosolutions, a biotech company working to develop pharmaceuticals to protect against bioweapons. Though his Immune System Support System enterprise failed within three years, the impression that this Russian-born U.S. citizen was trying to play the biosecurity game for his own profit cast a shadow on the altruistic motivations for which he was awarded the Barkley Medal for World Peace. See, for example, David Willman, "Selling the Threat of Bioterrorism," *Los Angeles Times*, July 1, 2007; Jeanne Lenzer, "Claim That Smallpox Vaccine Protects Against HIV is Premature, Say Critics," *British Medical Journal* 327, no. 7417 (2003): 699; "Out of the Box and Bottle: Biodefense Research, Remarkable Claims, Startling Patents, and Immune Systems Supplements," *National Security Notes*, March 31, 2006, www.globalsecurity.org/org/nsn/nsn-060331.htm, accessed August 10, 2016.

84. Butler and Spivak 2007, 4.

85. Masco 2008.

86. Masco 2010a, 140.

87. Masco argues that through wide-ranging practices of emotional management, the Cold War state established "logics of survival and sacrifice" by which individual desires and responses were subordinated to the preeminent goal of preserving collective life after nuclear catastrophe. Ibid, 150.

88. Friedberg 2000.

89. Masco 2008; Oakes 1994.

90. Grossman 2001, 12.

91. Enemark 2007. Mitchell Dean argues that risk is a way of presenting such events in order to make them governable, for there can be no security in a society that cannot calculate, and therefore cannot plan for, the worst imaginable disaster. Dean 2004, 183.

92. Collier, Lakoff, and Rabinow 2004.

93. Butler and Spivak 2007, 1.
94. See Oakes 1994, 6; Grossman 2001, xiii; Weart 1988; Masco 2008.
95. Weart 1988, 104.
96. Lakoff 2007.
97. Oakes 1994, 101.
98. Gray 1994.
99. Weart 1988.
100. Ewald 1993, 274.
101. As President Bush said on signing the bill, "The Act restructures and strengthens the executive branch of the Federal Government to better meet the threat to our homeland posed by terrorism." "President's Remarks at Homeland Security Bill Signing," White House, Office of the Press Secretary, November 25, 2002.
102. Homeland Security Act of 2002, 6 U.S.C. §111, Title I, section 101 b.1.a–f.
103. Donald F. Kettl, *The Department of Homeland Security's First Year: A Report Card* (New York: Century Foundation, 2004), 1.
104. This consolidation of power was rationalized as necessary to improve communication, coordination, and information gathering among the numerous executive agencies concerned with administering the nation's security systems. Campbell 1992.
105. 6 U.S.C. §121, Title II, section 201 d.2.
106. In turn, "The term 'assets' includes contracts, facilities, property, records, unobligated or unexpended balances of appropriations, and other funds or resources (other than personnel)." Homeland Security Act of 2002, 6 U.S.C. §101, sec. 2.3.
107. 6 U.S.C. §101, Introduction, sec. 2.9.
108. U.S. Department of Homeland Security, *Securing Our Homeland: U.S. Department of Homeland Security Strategic Plan* (Washington, DC: U.S. Department of Homeland Security, 2004).
109. U.S. Department of Homeland Security, "Ready.gov: Introduction; What Is www.ready.gov All About?" www.Ready.gov, February 19, 2003. www.geol.lsu.edu/hart/SURVIVAL/URBAN/fema02.htm, accessed August 10, 2016.
110. Wald 2002b.
111. Masco 2008.
112. Gregory Poland, "Bush's Bioshield Reignites Research," Minnesota Public Radio, January 30, 2003, http://news.mpr.org, accessed March 31, 2007.
113. *Subcommittee on Prevention of Nuclear and Biological Attack of the House Committee on Homeland Security: Engineering Bio-terror Agents: Lessons from the Offensive U.S. and Russian Biological Weapons Programs*, 109th Cong. (July 1, 2005) (statement of Roger Brent).
114. Ibid., 13.
115. U.S. Department of Homeland Security, *Securing Our Homeland*, 5.
116. President George W. Bush, July 21, 2004, cited in *Science and Technology: A Foundation for Homeland Security*, ed. Executive Office of the President (Washington, DC: Office of Science and Technology Policy, 2005), 1.

117. *Subcommittee on Prevention of Nuclear and Biological Attack*, statement of Roger Brent.

118. Braun and Whatmore 2010, xxi.

119. Ibid.

120. Bennett 2010, 96.

121. A similar question was posed in a hearing before the same committee eleven months later: "Mr. LINDER. Does it startle you—I may have asked you this the last time you were here—to know that significant numbers of Iranian children are being vaccinated for smallpox today?" *Reducing Nuclear and Biological Threats at the Source: Hearing Before the House Committee on Homeland Security*, 109th Cong., (June 22, 2006).

122. See for example, Neal Boortz, "Why are Iranian Children Vaccinated against Smallpox?" *Nealz Nuze, Free Republic* (blog), www.freerepublic.com/focus/f-news/1435455/posts, accessed April 1, 2007. No blogger identified a source for the speculation.

123. Masco 2010b, 461.

124. Jasanoff 2010, 15.

CHAPTER 2. MICROBES FOR WAR AND PEACE

1. Geissler and Moon 1999.

2. Protocol for the Prohibition of the Use in War of Asphyxiating, Poisonous or other Gases, and of Bacteriological Methods of Warfare (signed June 17, 1925, entered into force February 8, 1928), 94 League of Nations Treaty Series 66–74 (Geneva Protocol).

3. Merck 1946, 2.

4. Ibid., 4.

5. Ibid., 2.

6. Bennett 2010, xix.

7. Riley Housewright, "Riley Housewright Oral History Interview," December 12–13, 1995,13-II BP, folder 6, tape 3, page 9, Center for the History of Microbiology/American Society for Microbiology (CHOMA/ASM) Archives, University of Maryland Baltimore County, Baltimore, MD. Emphasis in original.

8. Fort Detrick. *Opportunity at Fort Detrick*, brochure, Frederick, MD, no date. Regional History, Maryland box, CHOMA/ASM Archives.

9. W.B. Dickinson Jr, "Government Research and Development: Efforts to Upgrade Armed Forces Laboratories," *Editorial Research Reports 1962*, vol. 1 (Washington, DC: CQ Press, 1962), http://library.cqpress.com/cqresearcher/cqresrre1962012400, accessed August 7, 2016.

10. Price 1965, 71.

11. Fothergill 1957, 865.

12. Williams 1980.

13. Bennett 2010, 12.

14. LeRoy D. Fothergill, "This Is the Biological Warfare Threat," paper presented at the Symposium on Non-military Defense, American Chemical Society, Cleveland, Ohio, April 8, 1960.

15. Housewright interview, 13-II, folder 6, tape 3, page 1.
16. Ibid.
17. Merck 1946.
18. Ibid., 5.
19. Ibid., 6.
20. Richard M. Clendenin, *Science and Technology at Fort Detrick, 1943–1968*, Silver Anniversary Publication, Technical Information Division, Fort Detrick, Frederick, MD, 1968, CHOMA/ASM Archives.
21. "Anniversary Edition: 20 Years for Fort Detrick," *News* (Frederick, MD), April 9, 1963, 9.
22. Wedum 1953, 1429.
23. Reitman 1956, 12.
24. Merck 1946 5.
25. Charles L. Baldwin and Robert S. Runkle, "Biohazards Symbol: Development of a Biological Hazards Warning Signal," paper presented at the 6th Annual Technical Meeting of the American Association for Contamination Control, Washington, DC, May 18, 1967. The paper also describes the work Baldwin did with Dow Chemical Company in the 1960s to create a symbol for biohazards which would be unambiguous and easily recognized. Baldwin's design continues to be used in laboratories and biological defense.
26. Greer Williams, "Laboratory against Death," *Cosmopolitan*, February 1947.
27. Wedum and Phillips 1964, 46.
28. Fothergill, "This is the Biological Warfare Threat," 4.
29. Druett, Henderson, Packman, and Peacock 1953; Harper and Morton 1953.
30. Fothergill, "This is the Biological Warfare Threat," 6.
31. Nomination for listing in National Register of Historic Places, One-Million-Liter Test Sphere (Horton Test Sphere), Fort Detrick, Frederick, Maryland, #77000696, http://msa.maryland.gov/megafile/msa/stagsere/se1/se5/010000/010400/010489/pdf/msa_se5_10489.pdf, accessed February 23, 2014.
32. Ibid., 4.
33. Fothergill 1964. *Bacillus atrophaeus*, also known as *B. subtilis* var. *niger* or *B. globigii* (BG), was cultivated at Fort Detrick for its aerobic spore-forming behavior similar to *B. anthracis*, or anthrax. The bacterium is harmless and heat tolerant, and produces a dark pigment that visually distinguishes it from other bacteria in its environment. Numerous strains descended from those developed at Detrick continue to be used in biosecurity research and other microbiological studies.
34. Fothergill, "This is the Biological Warfare Threat."
35. Fothergill 1964, 10.
36. Fothergill 1957, 20.
37. Williams, "Laboratory against Death."
38. They published their findings with other Detrick scientists in a series for the *American Journal of Medical Technology* in 1955 and 1956: Morton Reitman and G. Briggs Phillips, "Biological Hazards of Common Laboratory Procedures, I: The Pipette," *American Journal of Medical Technology* 21 (1955):

338–42; Everett Hanel Jr. and Robert Alg, "Biological Hazards of Common Laboratory Procedures, II: The Hypodermic Syringe and Needle," *American Journal of Medical Technology* 21 (1955): 343–46; Morton Reitman and G. Briggs Phillips, "Biological Hazards of Common Laboratory Procedures, III: The Centrifuge," *American Journal of Medical Technology* 22 (1956): 14–16; G. Briggs Phillips and Morton Reitman, "Biological Hazards of Common Laboratory Procedures, IV: The Inoculating Loop," *American Journal of Medical Technology* 22 (1956): 16–17.

39. Johansson and Ferris 1946.

40. Covert 1993.

41. Reitman and Wedum 1956, 664.

42. See, for example, Wedum et al. 1956.

43. Dick and Hanel 1970.

44. Reitman and Wedum 1956.

45. Wedum et al. 1956, 1109.

46. Jemski and Phillips 1963; Pike and Sulkin 1952.

47. Wedum et al. 1956, 1145.

48. Ibid.

49. Phillips et al. 1965, 16.

50. Jemski and Phillips, 1963, 10.

51. *Biohazard Control and Containment in Oncogenic Virus Research*, brochure, U.S. Department of Health, Education, and Welfare, National Institutes of Health, n.d. Laboratory Safety box, CHOMA/ASM Archives, 10.

52. Wedum and Phillips 1964, 50.

53. Wedum 1959, 105.

54. Wedum and Phillips 1964, 47.

55. Reitman and Wedum 1956, 664.

56. "Biohazard Containment and Control for Recombinant DNA Molecules: Fourth in a Series of Short Courses," course handout, University of Minnesota, School of Public Health, National Cancer Institute Office of Research Safety, September 7–8, 1977, n.p.

57. Wedum 1959, 119.

58. Richard M. Nixon, "Remarks Announcing Decisions on Chemical and Biological Defense Policies and Programs," White House press release, November 25, 1969. CHOMA/ASM Archives, 12-II BP, Folder 8.3.

59. Bingham and Hinchliffe, 2008.

CHAPTER 3. THE WILD MICROBIOLOGICAL WEST

1. Mary Wulff, conversation with author, July 9, 2008.

2. For explorations of this term, see Rose and Novas 2005. The definition claimed here builds on Adriana Petryna's expansion of the concept in *Life Exposed: Biological Citizens after Chernobyl*, 2002.

3. Rose and Novas 2005, 7.

4. Rose 2001.

5. In a laboratory, all spaces, practices, and materials are dedicated to the production of knowledge in the form of texts for publication. See Golinski

1998; Lynch 1995. Bruno Latour and Steve Woolgar (1986, 46) argue that this process leaves "inscriptions" on its outputs, the traces of actions in the laboratory that can be represented in publication, until they eventually strip scientific statements of their situated origins, thus producing "facts." With all its resources of material and energy, as well as devices of inscription, the laboratory gains authority over knowledge. Though nearly erased through the practices of publication, the materiality of the laboratory itself structures the work of knowledge making. The laboratory is a site of power, for it sustains the belief that truth exists outside its situated context and in turn explains what that truth is. As Donna Haraway observes (1991, 248 n. 2), "The laboratory for Latour is the railroad industry of epistemology, where facts can only be made to run on the tracks laid down from the laboratory out. Those who control the railroads control the surrounding territory."

6. Willy Burgdorfer, conversation with author, July 29, 2008.

7. When the building used by the lab prior to the 1930s was converted into a community playhouse in the 1980s, old specimen jars were found sitting on shelves, abandoned in a move half a century before. Disputes over landfill issues have popped up repeatedly in the lab's history, most recently regarding a dump site in Victor, Montana. The NIH agreed to contribute to the cleanup of this site.

8. Foucault (2007, 64) proposes that the effect of the security apparatus that emerges through control of epidemics hinges on the problem of circulation "in the very broadest sense of movement, exchange, and contact, as form of dispersion and also as form of distribution." See also Bingham and Hinchliffe 2008, 177.

9. "Darby Commercial Club to Battle Spotted Fever," *Western News*, February 14, 1913.

10. Montana State Board of Entomology, *Third Biennial Report of the Montana State Board of Entomology, 1917–18* (Helena, MT: Independent Pub. Co., 1918), 17.

11. Maier 2000.

12. Shared susceptibility to disease defines populations in terms of risk and establishes a level of action acceptable to the governing entity. See Foucault 2007.

13. Montana State Board of Entomology, *Fourth Biennial Report of the Montana State Board of Entomology, 1919–1920* (Helena, MT: Independent Pub. Co., 1921), 12.

14. Elsie McCormick, "Death in a Hard Shell," *Saturday Evening Post*, November 15, 1941, 24. Whether or not McClintic's death in the nation's capital was the event that finally pushed Rocky Mountain spotted fever to the national spotlight, the next year the federal government allocated fifteen thousand dollars to an antitick campaign in the Bitterroot Valley. That same year, the state of Montana established a board of entomology, appropriating an additional five thousand dollars "to investigate and study the dissemination by insects of diseases . . . having for its purpose the eradication and prevention of such diseases." Montana State Board of Entomology, *First Biennial Report of the Montana State Board of Entomology, 1913–1914* (Helena, MT: Independent Pub. Co., 1915), 5.

15. Montana State Board of Entomology, *Fourth Annual Report.* Control efforts did bring about results, reducing the number of spotted fever fatalities from eleven in 1913 to just three in 1918.

16. "Conference Is Held on Woodtick," *Western News,* April 15, 1913.

17. Montana State Board of Entomology, *First Biennial Report,* 18, 20. The recipe for the lethal water consisted of "8 or 8½ pounds of arsenite of soda (80% arsenious acid); 5½ pounds soft soap; 2 gallons paraffin (kerosene); 400 (Imp.) gallons of water (480 U.S. gal.)." *Stevensville Register,* May 22, 1913. Kerosene was used to "add penetration to the arsenic solution" and did not kill ticks. Montana State Board of Entomology, *First Biennial Report,* 16–17. When the mixture was not right, ranchers reported that the animals' skin was "somewhat burned or parched, and remain[ed] so for several days." *Stevensville Register,* May 22, 1913. Milk cows and workhorses were not subjected to the dip.

18. Montana State Board of Entomology, *First Biennial Report,* 5.

19. "Dipping Vat Destroyed by Vandals," *Western News,* Friday, June 13, 1913. Cement forms on each corner of the vat had been pried off, and the structure had been punctured, suggesting "that the damage had been done by a crowbar." Dunbar refused to pay one hundred dollars' bail, spending one night in jail before he was tried and found not guilty of the offense.

20. The report of the first year lists three strategies, with dipping at the top of the list. By 1919, however, the report showed seven components to the control plan. The story of the dipping vats and the dynamite is retold in the visitor center at RML as an example of the public opposition the lab "had to overcome" in its early history.

21. In 1915, workers applied one thousand pounds of strychnine-doused grain over one hundred square miles. They reported the cost of poisoning, including labor, at five hundred dollars for the season, less than the construction of a single dipping vat. Montana State Board of Entomology, *First Biennial Report,* 27. In 1912, the cost of a concrete dipping vat in the Bitterroot Valley, including corrals and dripping pens was about $520. Thomas B. McClintic, *Public Health Reports (1896–1970)* 27, no. 20 (May 17, 1912): 735.

22. McCormick, "Death in a Hard Shell," 24.

23. Montana State Board of Entomology, *Third Biennial Report,* 19.

24. Latour and Woolgar 1986, 51. At a conference in April 1923, Parker reported the incidence of the disease in ticks to be about 2 percent.

25. Philip 2000, 138.

26. Ibid., 140.

27. A 1926 leaflet calls the lab "anything but a safe place to rear the infected ticks," and a *Saturday Evening Post* article vividly describes the "dangerous work of tick rearing and vaccine making" in a tight, dark space, evoking language that resurfaced nearly a century later in protests against BSL4 facilities. McCormick, "Death in a Hard Shell," 24.

28. Montana State Legislature, House Bill No. 265, 20th Legislative Assembly, 1927.

29. The so-called Block 19 of the Pine Grove Addition was part of the Hamilton townsite but outside the incorporated city limits. *Court Transcript, Findings of Fact and Conclusions of Law, Section III.*

30. Record of the Fourth District Court, State of Montana, 1927, 8.
31. Wulff, conversation with author, July 9, 2008.
32. Jim Miller, conversation with author, July 19, 2008.
33. Petryna 2002, 6.
34. Ibid., 14.
35. One notable difference between the 1927 and 2004 lawsuits is that the threat of Rocky Mountain spotted fever used to rationalize the 1927 expansion was intimate and local, whereas the expansion in the twenty-first century was designed to deal with organisms that, by definition, are not localized. Though both eras were characterized by some popular concern over disease threats, the first expansion of RML brought microbes from across the river, whereas the BSL4 laboratory brought microbes that would otherwise enter the area only through human means. Still, both lawsuits were settled at the most local level by directing how the laboratory space would be managed for the security of local citizens, a material outcome showing the sway of the community in regulating laboratory space.
36. Record of the Fourth District Court, State of Montana, 1927. The speaker is citing the state constitution, paraphrasing section 3, "Inalienable Rights."
37. Ibid., opening statement, 4.
38. Ibid., 41–42.
39. Ibid., 54–55.
40. Ibid., opening statement, 4.
41. Ibid., 6.
42. Ibid. Mary Wulff later expressed frustration at the scientists' insistence that the building was completely secure. "If they would just admit that something bad might happen, we would have some common ground. But they just kept saying over and over again that there was *no* risk." Wulff, conversation with author, July 9, 2008.
43. Marshall Bloom, conversation with author, July 17, 2008. John Swanson believes such protocols came about once the diseases being studied were deadly enough that a needle prick could be a death sentence, but also as social concern increased, not just about disease but also about laboratory practice. Swanson, conversation with author, July 10, 2008.
44. Record of the Fourth District Court, State of Montana, 1927, xvi.
45. Ibid., 3–7.
46. Philip 2000, 171.
47. Maurine Hughes, conversation with author, July 18, 2008; Bill Hadlow, conversation with author, July 15, 2008.
48. Burgdorfer, conversation with author, July 29, 2008.
49. The Walsh Act appropriated $150,000 for the laboratory, part of which was used to purchase Building 1, and the remainder of which paid for the construction of Building 2. An allotment in 1935 paid $132,000 for Building 4 and two residences. A further $622,000 was allocated in 1938 for Building 3 and three other buildings.
50. Larry D. Swanson, *2002 Ravalli County Economic Needs Assessment: The Bitterroot Valley Economy*, prepared for the Ravalli County Economic Development Authority, November 2002.

51. Ravalli County Economic Development Authority, *What is RCEDA?*, brochure, 2008. RCEDA is a Port Authority with 501(c)(3) tax status, funded by grants from the county and Hamilton City as well as private donations.

52. Most rural areas would not be able to raise sufficient matching funds to access these grants; however, because the land donation was appraised according to high land values in the area, the community raised $1.6 million to match the grant.

53. In July 2008, GSK laid off fourteen people, bringing the total number of employees from 292 to 278. At the time, Ken Meyers cited 260 as the steady number of employees at GSK. The $3 million figure is Foster's estimate. In 2015, GSK announced it would be closing the vaccine research branch and laying off 27 of the 211 employees in the Hamilton facility. Perry Backus, "GlaxoSmithKline to lay off 27 from Hamilton operations," *Missoulian*, Febuary 6, 2015. http://missoulian.com/news/state-and-regional/glaxosmithkline-to-lay-off-from-hamilton-operations/article_48f3ff1b-49f5–56d5–9e58-a3fe1e29497a.html, accessed August 7, 2016.

54. Kaushik Sunder Rajan (2006) argues that entirely new forms of capital emerge alongside the new biology, commodifying life itself.

55. Rabinow 1996, 99.

56. Rose and Novas 2005, 7.

57. Petryna 2002.

58. Rabinow 1996, 98.

59. Petryna 2002, 14–15. See also Nguyen 2005; Rose and Novas 2005.

60. Jasanoff 2010, 30, 36.

61. Though the protest in Hamilton did not impede the BSL4 expansion, other communities have successfully intervened in the establishment of biolabs. Thomas Beamish (2015) compares three community protests of biodefense laboratories, showing how deep social differences were overcome in alliances against the laboratory.

62. Alex Gorman, conversation with author, July 21, 2008.

63. Jenny Johnson, "Lab to Play Expanded Role in Fighting Bioterrorism," *Ravalli Republic*, February 11, 2002, http://ravallirepublic.com/articles/2002/02/11/news/export3133.txt, accessed July 14, 2008.

64. Gorman, conversation with author, July 21, 2008.

65. "NIH Director Speaks to Community," *Ravalli Republic*, September 15, 2002, http://ravallirepublic.com/articles/2002/09/15/news/export4258.txt, accessed July 18, 2008.

66. Swanson, conversation with author, July 10, 2008.

67. Bloom, conversation with author, July 17, 2008.

68. Notably, the CLG was formed after the decision had been made to proceed with the EIS. Only then could the laboratory officially acknowledge that its actions had effects beyond its boundaries without undermining the arguments of the initial environmental assessment.

69. NEPA's scoping process involves a comment period that helps to define the scope. Scope includes actions (connected, cumulative, and similar), alternatives (no action, other reasonable action, and mitigation), and impacts (direct, indirect, and cumulative).

70. During the Draft EIS scoping process, 588 public comments were received in 103 separate documents. They focused primarily on alternatives, mitigation, and effects analysis. National Institutes of Health, *Draft Final Environmental Impact Statement, RML Integrated Research Facility, May 2003*, 1.8–1.9. All comments can be found in National Institutes of Health, *Final Environmental Impact Statement, Rocky Mountain Laboratories Integrated Research Facility, Response to Comments*, April 2004, 5.1–5.282.

71. Another innovative proposal to ensure the security of the community was that the names of all agents being studied in the labs be distributed to the local doctors who would be diagnosing diseases in the citizen population.

72. Giddens 1999, 8, 3.

73. Giddens 1999, 4.

74. Science takes place in a political context in which it is both the subject and the object of power relations. As Petryna writes (2002, 10), "The processes of making scientific knowledge are inextricable from the forms of power those processes legitimate and even provide solutions for." In the crisis of bioterrorism, cultural fears raise the desire for knowledge, demanding science to simultaneously produce both evidence of and solutions to the crisis.

75. National Institutes of Health, *Draft Environmental Impact Statement*, 2.11, 4.2.

76. Ibid., 4.2, emphasis added.

77. Ibid., 6.4.4c. The comments submitted by the advocacy groups cited other EIS documents that preliminarily found no risk and yet still followed up these assessments with study. For example, a study of brucellosis in bison stated that there were no known cases of transmitting the disease between these species, yet they still studied all the literature that had led to this conclusion. No risk, in this case, did not absolve the drafters of the EIS from the responsibility to study the risk.

78. Dean 2004, 178.

79. The question follows of why citizens themselves cannot quantify the risk. Several circumstances sustain the authority of the state over the calculation of risk. First, through the federally mandated NEPA process, citizens transfer responsibility for risk assessment to an entity that has the skills and resources to gather scientific information on a topic. The public process also presumes that determinations of risk have material outcomes in the capitalist quest for economic advancement and that entities might deliberately distort dangers in order to achieve independent objectives. Finally, because of the security practices of the science complex, citizens do not feel they have access to the information necessary to conduct a valid and complete risk assessment.

80. Gorman, conversation with author, July 21, 2008.

81. National Institutes of Health, *Final Environmental Impact Statement, Rocky Mountain Laboratories Integrated Research Facility* (FEIS), July 26, 2004, 4.11.

82. Ibid., 4.12, 4.13.

83. This phrase is repeated three times in the document, the second and third times worded as "not a significant risk." Ibid., S.4, 4.7. The difference between no risk and insignificant risk was important in public attempts to understand

the risk posed by the IRF. The repetition of the phrase underscores links between safe behavior and safe laboratories, as well as past safety and future safety.

84. Larry Campbell, conversation with author, July 12, 2008.

85. Ibid.

86. Transcribed from video of the DEIS scoping meeting, September 18, 2002, accessed at Rocky Mountain Laboratories library.

87. Giddens 1999, 5.

88. Petryna 2002, 63.

89. Council on Environmental Quality, Regulations for Implementing NEPA, Section 1502.2, Implementation, *Code of Federal Regulations*, U.S.C. 40 CFR 1502.2 (g).

90. Gorman, conversation with author, July 21, 2008. See also also Jenny Johnson, "Lab to Play Expanded Role Fighting Bioterrorism," *Ravalli Republic*, February 11, 2002.

91. Rose 2001, 21.

92. When confronted with the memo, NIH officials accused opponents of the lab expansion of writing the memo to support some sort of conspiracy theory.

93. The memo is titled "Construction and Operation of an Infectious Disease Biosafety Level-4 Building at NIH, NIAID, DIR, Hamilton, Montana Campus," date-stamped "12/15/2000 5:17 P.M." It was released to the public by NIH on January 9, 2003, in response to FOI Case No. 27890, December 2000. In the Freedom of Information Act request, the FOI coordinator says, "Although it does not fit specifically under any of the questions you asked, included also is a copy of a memo, developed in December 2000 by the Director of NIAID's Division of Intramural Research, expressing the need for the construction of a BSL-4 building at the Rocky Mountain Laboratories location." Ironically, the controversial memo worked in the Boston activists' favor, because it argued that such labs should not be built in populous areas like Boston but in rural areas like Hamilton. Adam Smith, "NIAID Memo Suggests Level 4 Biodefense Labs Better in Unpopulated Areas to Avoid Major Public Health Disaster," *Nukewatch: Nuclear Watch New Mexico*, January 16, 2004, www.nukewatch.org/media2/postData.php?id=699, accessed August 7, 2016.

94. Campbell, conversation with author, July 12, 2008.

95. Ibid.

96. Gorman, conversation with author, July 21, 2008.

97. Indeed, part of the settlement agreement stipulates that NIH "will not weaponize any pathogen at its facility." The settlement agreement defines weaponization as "the manipulation of pathogens to render them more useful as weapons. NIH may study pathogens that have been weaponized." Settlement Agreement, Coalition for a Safe Lab, et al., v. National Institutes of Health, et al., in the United States District Court for the District of Montana Missoula Division, CV 04–158-M-DWM, September 27, 2004.

98. "Tick Warriors," *Saturday Evening Post*, November 15, 1951.

99. Settlement Agreement, 3.

100. Lakoff 2010, 5.

101. Petryna 2002, 218.

102. Rose 2001, 6.

103. Rabinow 2005, 43.

CHAPTER 4. AGENTS OF CARE

1. In a 2009 Gallup poll, 61 percent of Americans ranked the CDC as doing an excellent or good job, the highest score of any agency. Lydia Saad, "CDC Tops Agency Ratings, Federal Reserve Board Lowest," Gallup, July 27, 2009. www.gallup.com/poll/121886/cdc-tops-agency-ratings-federal-reserve-board-lowest.aspx, accessed November 8, 2009.

2. Foucault 2007.

3. Tomes 1999.

4. Graham, Talent, and Allison 2008, 23.

5. Centers for Disease Control and Prevention, "Centers for Disease Control and Prevention (C)," www.cdc.gov/maso/pdf/cdcmiss.pdf, accessed August 3, 2016.

6. U.S. Congress, House, Subcommittee on Prevention of Nuclear and Biological Attack, Committee on Homeland Security, *Implementing the National Defense Strategy*, 119th Congress (statement of Julie Gerberding, CDC director), July 28, 2005. Statement available at www.hhs.gov/asl/testify/t050728a.html, accessed August 8, 2016.

7. CDC also manages funds for external programs, such as children's vaccines and World Trade Center health recovery. With these appropriations, CDC's expenses in 2012 totaled $11,196,121,000. The figure of 23 percent reflects the actual budget authority of CDC. Budgets at www.cdc.gov/fmo/topic/budget%20information/index.html, accessed October 6, 2012 When President Obama appointed Thomas Frieden as CDC director, Frieden dissolved the unpopular structure of coordinating offices at CDC, reportedly to reduce layers of management. With no substantial changes in function, the Coordinating Office for Terrorism Preparedness and Emergency Response (COTPER) is now the Office of Public Health Preparedness and Response (PHPR). In Fiscal Year 2015, PHPR was funded at $1.7 billion. Centers for Disease Control and Prevention, *FY 2017 President's Budget Request Fact Sheet*, n.d., 2, www.cdc.gov/budget/documents/fy2017/phpr-factsheet.pdf, accessed August 7, 2016. That year, CDC's total operating budget was $8.8 billion which included $1.77 billion for Ebola-related activities and $15 million for Influenza program activities. Centers for Disease Control and Prevention, *Fiscal Year 2015 Annual Report*, Office of Financial Resources, n.d., 17, www.cdc.gov/fmo/topic/budget%20information/index.html, accessed August 31, 2016.

8. Centers for Disease Control and Prevention, *Budget Request Summary Fiscal Year 2007*, February 2006, https://stacks.cdc.gov/view/cdc/27906/Share, accessed August 7, 2016.

9. U.S. Congress, *Implementing the National Defense Strategy*.

10. Rabinow 1996, 98.

11. E. Cohen 2009, 63.

12. Scholars have also argued that the germ theory rationalized the victimization of certain populations through public health measures, such as steriliza-

tion and mass quarantine, and that the cultural context of war sustained a number of practices that would otherwise not be tolerated in society. See for example, Weindling 2000; Maclean 2008; Armstrong 1995.

13. See Foucault 2007; Rose 2007.

14. Foucault 2007, 59.

15. See Lakoff 2008; Collier 2008; Lakoff and Collier 2008.

16. Lakoff and Collier 2008, 17.

17. Pamela Diaz, conversation with author, November 18, 2008.

18. Dennis O'Mara, conversation with author, August 19, 2008.

19. Dixie Snider, conversation with author, November 18, 2008.

20. Centers for Disease Control and Prevention (CDC), *Advancing the Nation's Health: A Guide to Public Health Research Needs, 2006–2015* (Atlanta: CDC, 2006), 1, 35.

21. Shelton et al. 2012, 2759.

22. CDC, *Advancing the Nation's Health*, 36.

23. Brier 2011.

24. Diaz, conversation with author, November 18, 2008.

25. Stephen Morse, conversation with author, November 20, 2008.

26. Centers for Disease Control and Prevention, *2015: National Snapshot of Public Health Preparedness*, www.cdc.gov/phpr/pubs-links/2015/documents/2015_Preparedness_Report.pdf, accessed February 25, 2016.

27. CDC, *Advancing the Nation's Health*, 41.

28. Myers 2009.

29. Wald 2008, 71.

30. Foege 2011, 75.

31. Haggerty 2006.

32. Foucault 2007.

33. On the National Syndromic Surveillance Program and BioSense 2.0, see the CDC's website, www.cdc.gov/nssp/biosense/, accessed August 7, 2016. The BioSense Community Forum and other information for users is at sites.google.com/site/biosenseredesign, accessed August 7, 2016.

34. For a thorough discussion of local responses to BioSense, see Fearnley 2008.

35. Snider, conversation with author, November 18, 2008.

36. Foucault 2007.

37. Dan Sosin, conversation with author, December 15, 2008.

38. Lyon 2003. See also Bigo 2006.

39. CDC, *2015: National Snapshot of Public Health Preparedness*, 13.

40. Graham, Talent, and Allison 2008, xiii.

41. From 2007 to 2008, CDC.gov received an average of 41 million hits per month, with about 2.2 million searches run on the site each month. *CDC Fact Sheet*, www.cdc.gov/Other/pdf/CDCFactSheet.pdf, accessed March 27, 2009.

42. Snider, conversation with author, November 18, 2008.

43. Jim Curran, conversation with author, November 19, 2008.

44. Andy Mullins and Jim Hayslett, slides from a PowerPoint presentation given during EIS training, Spring 2004, CDC Library, Atlanta, Georgia.

45. Von Roebuck, conversation with author, November 20, 2008.

46. Department of Health and Human Services and Centers for Disease Control and Prevention, "Interim Pre-pandemic Planning Guidance: Community Strategy for Pandemic Influenza Mitigation in the US," released February 2007, www.flu.gov/planning-preparedness/community/community_mitigation.pdf, accessed August 5, 2016.

47. Bartlett and Borio 2008, 919.

48. Lisa Rotz, conversation with author, November 19, 2008.

49. Ong 2006, 6.

50. Susan True, conversation with author, November 21, 2008.

51. CDC, *2015: National Snapshot of Public Health Preparedness*, 7.

52. Ralph O'Connor, conversation with author, November 10, 2008.

53. Rotz, conversation with author, November 19, 2008; Diaz, conversation with author, November 18, 2008; Morse, conversation with author, November 20, 2008.

54. True, conversation with author, November 21, 2008.

55. CDC Foundation, "Meta-leadership Summit," www.cdcfoundation.org/programs/CDCFoundationInitiatives.aspx, accessed May 25, 2009.

56. CDC Foundation, 2008 *Report to Contributors*, www.cdcfoundation.org/sitefiles/ReportToContributors_FY08.pdf, accessed May 24, 2009. The fact that the contributors include biotechnology corporations poses potential conflicts of interest.

57. Kevin Brady, conversation with author, November 21, 2008.

58. Sosin, conversation with author, December 15, 2008.

59. CDC, *2015: National Snapshot of Public Health Preparedness*, iii.

60. Ibid., 17.

61. Ibid., 174.

CHAPTER 5. SIMULATION SCIENCE

1. 149 Cong. Rec. S9787 (2003) (statement of Senator Pete Domenici of New Mexico regarding the importance of the National Infrastructure Simulation and Analysis Center and the Emergency Response Training, Research and Development Center in Playas, New Mexico).

2. Weart 1988.

3. Masco 2008, 362.

4. Lakoff 2007, 1.

5. Collier 2008, 244.

6. Ibid.

7. Bacon 1989, 16; Horkheimer and Adorno 2007, 31.

8. Leiss 2007.

9. Margulis 1998, 84.

10. Wald, 2012.

11. Committee on Smallpox Vaccination Program Implementation, *Review of the Centers for Disease Control and Prevention's Smallpox Vaccination Program Implementation: Letter Report #6* (Washington, DC: National Academies Press, 2004), 21.

12. Taylor 2003, 8.

13. Teclaw 1979, 103.
14. Gusterson 2008, 559.
15. Gusterson 2008.
16. Taylor 2003, 9.
17. McKenzie 2004, 2044.
18. See, for example, Samuel A. Bozzette, Rob Boer, Vibha Bhatnagar, Jennifer L. Brower, Emmett B. Keeler, Sally C. Morton, and Michael A. Stoto, "A Model for a Smallpox-Vaccination Policy," *New England Journal of Medicine* 348, no. 5 (2003): 416–25; Martin Eichner and Klaus Dietz, "Transmission Potential of Smallpox: Estimates Based on Detailed Data from an Outbreak," *American Journal of Epidemiology* 158 (2003): 110–17; Raymond Gani and Steve Leach, "Transmission Potential of Smallpox in Contemporary Populations [Letter]," *Nature* 414, no. 6865 (2001): 748–51; Monica Giovachino, "Modeling the Consequences of Bioterrorism Response," *Military Medicine* 166, no. 11 (2001): 925–30; M. Elizabeth Halloran, Ira M. Longini Jr., Azhar Nizam, and Yang Yang, "Containing Bioterrorist Smallpox," *Science* 298 (2002): 1428–32; Nathaniel Hupert, Alvin I. Mushlin, and Mark A. Callahan, "Modeling the Public Health Response to Bioterrorism: Using Discrete Event Simulation to Design Antibiotic Distribution Centers," *Medical Decision Making* 22 (October 2002): S17–S25; Edward H. Kaplan, David L. Craft, and Lawrence M. Wein, "Analyzing Bioterror Response Logistics: The Case of Smallpox," *Mathematical Biosciences* 185 (2003): 33–72; Martin I. Meltzer, Inger Damon, James W. LeDuc, and J. Donald Millar, "Modeling Potential Responses to Smallpox as a Bioterrorist Weapon," *Emerging Infectious Diseases* 7, no. 6 (2001): 959–69; Martin I. Meltzer, "Risks and Benefits of Preexposure and Postexposure Smallpox Vaccination," *Emerging Infectious Diseases* 9, no. 11 (2003): 1363–70.
19. McKenzie 2004.
20. Epstein et al. 2002, 2.
21. Molina 2006, 2.
22. McKenzie 2004, 2046.
23. Certainly some models address mutation and change in microbes as well, but because preparedness models seek to identify places where human behavior can shape biological outcomes, they tend to make those components the variables, rather than the behavior of the microorganism.
24. Turkle 2009, 81.
25. Committee on Smallpox Vaccination Program Implementation, *Review.*
26. Turkle 2009, 7.
27. Gusterson 2008, 551.
28. Epstein et al. 2002, 1.
29. Ferguson et al. 2003, 681.
30. Committee on Smallpox Vaccination Program Implementation, *Review,* 22.
31. Federal Emergency Management Agency, "Fact Sheet: National Exercise Program (NEP), National Level Exercise—Capstone Exercise 2014," www.fema.gov/media-library-data/1391701556671-2204c5ec1c30a48dd0b783989206b68/nep.pdf, accessed August 9, 2016.

32. Nina Berman, conversation with author, July 29, 2009.

33. Allen Pitts in "Connecticut ARES TOPOFF 3 After Action Report," December 20, 2005, www.ctares.org/CT%20ARES%20TO3%20AAR%20FINAL.pdf, September 4, 2010.

34. See Wald 2012.

35. Inglesby, Grossman, and O'Toole 2001.

36. Turkle 2009, 8.

37. Myers 2009.

38. Committee on Smallpox Vaccination Program Implementation, *Review*, 30.

39. Weart 1988.

40. Ibid.

41. Oakes 1994; Masco 2008.

42. Lakoff and Collier 2010.

43. Committee on Smallpox Vaccination Program Implementation, *Review*, 22.

44. Department of Homeland Security, "TOPOFF: Exercising National Preparedness," www.dhs.gov/files/training/gc_1179350946764.shtm, accessed October 11, 2009.

45. Federal Emergency Management Agency, "Fact Sheet."

46. Federal Emergency Management Agency, *After Action Quick Look Report, TOPOFF 4 Full-Scale Exercise*, November 19, 2007, https://training.fema.gov/hiedu/docs/topoff4_afteraction_report2007.pdf, accessed Aug 9, 2016.

47. Simpson 2002, 56.

48. Inglesby, O'Toole, and Grossman 2001, 436.

49. Samimian-Darash 2009, 479.

50. Lakoff 2008b, 35.

51. De Goede 2008, 171.

52. Liane Hansen, "Is the U.S. Prepared for the Next Disaster?," interview with Craig Fugate, *Weekend Edition*, NPR, March 20, 2011. Transcript at www.npr.org/2011/03/20/134706175/Is-The-U-S-Prepared-For-The-Next-Disaster, accessed August 8, 2016.

53. For a detailed report on Dark Winter, see Roman 2002 and the CD-ROM published by the Center for Strategic and International Studies (Washington, DC: CSIS, 2001). See also O'Toole, Mair, and Inglesby 2002. Records of Atlantic Storm are archived at www.atlantic-storm.org.

54. Bingham and Hinchliffe 2008, 177.

55. Koh et al. 2008.

56. Greg Sanchez, personal communication, July 21, 2009.

57. Like the CRI, Operation Stonegarden channels funding to communities on the U.S. border with an objective of increasing border security, while the Urban Area Security Initiative directs money to cities with dense populations. A series of Border Health Security Acts have proposed expanding allocations to communities through federal grant programs. Though a version of this legislation has been introduced in every Congress since 9/11 save two, it has never been signed into law.

58. See for example Kuletz 1998; Voyles 2015.
59. Phelps-Dodge Corporation, *Annual Report*, 1971.
60. James W. Harrison, *Comprehensive Plan for Development, Hidalgo County, New Mexico*, Southwest New Mexico Council of Governments, May 1974.
61. Mark Vitalia, conversation with author, July 28, 2009; Anneliese Kvamme, conversation with author, April 14, 2010.
62. Environmental Protection Agency, *Economic Impacts of Air Emission Standards: Primary Copper Smelting*, 1997, www.epa.gov/ttnecas1/regdata/IPs/Primary%20Copper%20Smelting_IP.pdf, accessed July 24, 2009.
63. Wall display at the museum of the Lordsburg Historical Society, Lordsburg, NM, April 2010.
64. "Hidalgo Medical Services, Lordsburg, New Mexico," *Innovations* 1 (2004): 6.
65. George Zamora, "NM Tech Hosts Inaugural Event at Playas Training Center," New Mexico Tech news web page, December 2, 2004, www.nmt.edu/nmt-news/97–2004/2714–2dec01g, accessed August 8, 2016.
66. In 2010 New Mexico Senator Tom Udall requested a $30 million appropriation from the Appropriations Subcommittee on Homeland Security for "National Domestic Preparedness Consortium, New Mexico Institute of Mining and Technology, WMD First Responder Training Programs." For state appropriations, see New Mexico Legislature. Senate. S.B. 465, 47th Leg., 1st sess. (2005); New Mexico Legislature. House. H.B. 523, 47th Leg., 2nd sess. (2006); New Mexico Legislature. House. H.B. 622, 47th Leg., 2nd sess. (2006).
67. "NM Tech Lands $27.5 Million Deal," *Albuquerque Journal*, March 15, 2011.
68. Kvamme, conversation with author, April 14, 2010.
69. Erickson and Barratt 2004.
70. Der Derian 2001, 9.
71. Geyer 1989, 75–80.
72. Masco 2006, 206.
73. See, for example, Eric Lipton, "Come One, Come All, Join the Terror Target List," *New York Times*, July 12, 2006.

CHAPTER 6. BIOTERROR BORDERLANDS

1. Wald 2002b, 623.
2. Massumi 1993.
3. David V. Holtby, "World War I and the Federal Presence in New Mexico: The Punitive Expedition and the Education of General John J. Pershing," Centennial of New Mexico Statehood collection, University of New Mexico Repository, URI, http://repository.unm.edu/handle/1928/6708, 2008.
4. Stern 1999, 61–62.
5. Ibid.
6. Molina 2006, 195.
7. Stern 1999, 48.
8. Ibid., 52.

9. Ibid., 68.

10. Annette Anderson, conversation with author, July 27, 2009.

11. Rick Hampson, "Pancho Villa Now Celebrated in New Mexico," *USA Today*, March 9, 2011.

12. Anderson, conversation with author, March 12, 2011.

13. New Mexico Energy, Minerals, and Natural Resources Department, "Pancho Villa State Park and Village of Columbus Celebrate Camp Furlong Day," press release, March 2, 2011, www.emnrd.state.nm.us/PRD/documents/PR-ParksPanchoVillaStateParkCelebratesCampFurlongDayMarch12_2011.pdf, accessed March 15, 2011.

14. New Mexico Border Authority, *Livestock*, www.nmborder.com/Livestock.aspx, accessed August 7, 2016.

15. Bruce Whetten, "Local Gun Sales on the Rise," *Douglas Dispatch*, April 7, 2010.

16. Appadurai 2006, 99, 101, 104.

17. New Mexico Department of Agriculture, *NMDA 2007–2009 Biennial Report*, http://nmdaweb.nmsu.edu/events-and-publications-folder/Binder1.pdf, accessed November 10, 2010.

18. In 2014, the SWBFSDC changed its name to the Southwest Border Food Protection and Emergency Preparedness Center.

19. Billie Dictson, conversation with author, April 19, 2010.

20. See the website of the Customs and Border Patrol at www.cbp.gov/newsroom, accessed March 3, 2011.

21. U.S. Congress, House, Committee on House Homeland Security, Subcommittee on Emergency Communications, Preparedness, and Response, *Improving Emergency Preparedness and Response Capabilities*, 110th Cong., July 19, 2007 (statement of Alfonso Martinez-Fonts Jr., assistant secretary, Private Sector Office, Office of Policy, Office of the Secretary, Department of Homeland Security).

22. Billie Dictson and Jeff Witte, conversation with author, April 19, 2010.

23. Bingham and Hinchliffe 2008, 177.

24. Dictson, conversation with author, April 19, 2010.

25. *Homeland Security Presidential Directive/HSPD-5—Management of Domestic Incidents*, February 28, 2003, 229, www.gpo.gov/fdsys/pkg/PPP-2003-book1/pdf/PPP-2003-book1-doc-pg229.pdf, accessed July 3, 2014.

26. Ibid.

27. Federal Emergency Management Authority, *The FEMA Training Module 103 for Incident Command Systems*, www.fema.gov/about/training/emergency.shtm, accessed June 14, 2010.

28. *Homeland Security Presidential Directive/HSPD-5*, 229.

29. Dictson, conversation with author, April 19, 2010.

30. Muntean 2009.

31. Butler 2008, 923.

32. Richard P. Wenzel, "What We Learned from H1N1's First Year," *New York Times*, April 13, 2010, www.nytimes.com/2010/04/13/opinion/13wenzel.html. Natalia Molina writes of similar anxieties that beset Los Angeles in the 1930s. One white doctor predicted: "Without adequate isolation quarantine

hospital facilities [*sic*], Los Angeles would be subject to all the adverse publicity that such a condition would occasion, with resultant loss of millions in hotel and tourist trade and curtailment of general business activities." Molina 2006, 130.

33. Choffnes, Mack, and Relman, 2010.

34. Douglas A. McIntyre, "Can the Epidemic Take Down Global Markets?" *Newsweek*, April 27, 2009, www.newsweek.com/id/195221, accessed April 27, 2009.

35. Centers for Disease Control and Prevention, *The 2009 H1N1 Pandemic: Summary Highlights, April 2009–April 2010*, www.cdc.gov/h1n1flu/cdcresponse.htm, accessed August 4, 2016.

36. Gibbs, Armstrong, and Downie 2009, 207.

37. Rob Smith, "Border Crossing!," cartoon, *Rob's Right: Political Commentary From a Conservative Cartoonist*, www.robsright.com/tag/swine-flu/, May 1, 2009, accessed August 7, 2016.

38. Ed Stein, "Swine Flu," cartoon, *Amite-Tangi Digest*, May 1, 2009, www.amitetoday.com/swine-flu-cartoon, accessed May 4, 2009.

39. "The New Swine Flu," editorial, *New York Times*, April 28, 2009.

CONCLUSION

1. Dennis P. Merklinhaus, "Freaked Out Yet?" *Military Technology* 8, no. 11 (2014): 1.

2. Boddie et al. 2015.

3. A 2015 study further suggests that humans have acquired microbes' genes through horizontal gene transfer, meaning that human DNA may contain microbial DNA. Crisp et al. 2015.

4. Ilchmann and Revill 2014.

Selected Bibliography

I have listed below selected works that influenced the writing of this book. The sources that help to tell the stories of bioterrorism as it emerged in the communities I visited, including interviews, news reports, and government documents, are cited only in the backnotes.

Agamben, Giorgio. 1998. *Homo Sacer: Sovereign Power and Bare Life*. Stanford, CA: Stanford University Press.

———. 2000. *Remnants of Auschwitz: The Witness and the Archive*. New York: Zone Books.

———. 2001. "On Security and Terror." *Frankfurter Allgemeine Zeitung*, September 20.

———. 2005. "The State of Exception as a Paradigm of Government." In *State of Exception*, 1–31. Chicago: University of Chicago Press.

Amass, Sandra F., ed. 2006. *The Science of Homeland Security*. West Lafayette, IN: Purdue University Press.

Anderson, Fred. 2000. *Crucible of War: The Seven Years' War and the Fate of Empire in British North America, 1754–1766*. New York: Alfred A. Knopf.

Appadurai, Arjun. 1996. *Modernity at Large: Cultural Dimensions of Globalization*. Minneapolis: University of Minnesota Press.

———. 2006. *Fear of Small Numbers: An Essay on the Geography of Anger*. Durham, NC: Duke University Press.

Armstrong, David. 1995. "The Rise of Surveillance Medicine." *Sociology of Health and Illness* 17(3): 393–404.

Bacon, Francis. 1989. *New Atlantis and the Great Instauration*. Edited by Jerry Weinberger. Wheeling, IL: Crofts Classics.

Baer, Martha. 2005. *Safe: The Race to Protect Ourselves in a Newly Dangerous World*. New York: HarperCollins.

Baldwin, Charles L., and Robert S. Runkle. 1967. "Biohazards Symbol: Development of a Biological Hazards Warning Signal." Paper presented at the 6th Annual Technical Meeting of the American Association for Contamination Control, Washington, DC, May 18.

Barnett, Michael N., and Thomas George Weiss. 2008. *Humanitarianism in Question: Politics, Power, Ethics*. Ithaca, NY: Cornell University Press.

Barras, Vincent, and Gilbert Greub. 2014. "History of Biological Warfare and Bioterrorism." *Clinical Microbiology and Infection* 20(6): 497–502.

Barrett, Ronald. 2006. "Dark Winter and the Spring of 1972: Deflecting the Social Lessons of Smallpox." *Medical Anthropology* 25: 171–91.

Bartlett, John G., and Luciana Borio. 2008. "Healthcare Epidemiology: The Current Status of Planning for Pandemic Influenza and Implications for Health Care Planning in the United States." *Clinical Infectious Diseases* 46(6): 919–25.

Baudrillard, Jean. 1998. "The End of the Millennium or the Countdown." *Theory, Culture and Society* 15(1): 1–9.

Baudrillard, Jean, and Michel Valentin. 2006. "War Porn." *Journal of Visual Culture* 5(1): 86–88.

Bauman, Zygmunt. 2003. "In and Out of the Toolbox of Sociality." In *Liquid Love: On the Frailty of Human Bonds*, 38–76. Hoboken, NJ: Wiley.

———. 2006. *Liquid Fear*. Cambridge: Polity Press.

———. 2007. *Liquid Times: Living in an Age of Uncertainty*. Cambridge: Polity Press.

Beamish, Thomas D. 2015. *Community at Risk: Biodefense and the Collective Search for Security*. Stanford, CA: Stanford University Press.

Beck, Ulrich. 1992. *Risk Society*. Translated by Mark Ritter. London: Sage.

Behbehani, Abbas M. 1980. "Decline and Fall of the Smallpox Empire." *World Health: The Magazine of the World Health Organization*. May.

———. 1983. "The Smallpox Story: Life and Death of an Old Disease." *Microbiology Review* 47(4): 455–509.

Bennett, Jane. 2010. *Vibrant Matter: A Political Ecology of Things*. Durham, NC: Duke University Press.

Bentham, Jeremy. 1871. *Theory of Legislation*. Edited by Étienne Dumont. Translated by Richard Hildreth. London: Trübner.

Bigo, Didier. 2002. "Security and Immigration: Toward a Critique of the Governmentality of Unease." *Alternatives: Global, Local, Political* 27(1): 63.

Bingham, Nick, and Steve Hinchliffe. 2008. "Mapping the Multiplicities of Biosecurity." In *Biosecurity Interventions: Global Health and Security in Question*, edited by Andrew Lakoff and Stephen J. Collier, 173–94. New York: Columbia University Press.

Bishop, George D. 1991. "Lay Disease Representations and Responses to Victims of Disease." *Basic and Applied Social Psychology* 12(1): 115–32.

Blaine, Harden. 2000. "Ebola's Shadow." *New York Times Magazine*. December 24, 44.

Blank, Andreas. 1999. "Why Do New Meanings Occur? A Cognitive Typology of the Motivations for Lexical Semantic Change." In *Historical Semantics and Cognition*, edited by Andreas Blank and Peter Koch, 61–90. Berlin: Mouton de Gruyter.

Blum, Rony. 2008. "'Ethnic Cleansing' Bleaches the Atrocities of Genocide." *European Journal of Public Health* 18(2): 204–9.

Boddie, Crystal, Matthew Watson, Gary Ackerman, and Gigi Kwik Gronvall. 2015. "Assessing the Bioweapons Threat: Is There a Foundation of Agreement among the Experts?" *Science*, August 21, 792–93.

Bourke, Joanna. 2006. *Fear: A Cultural History*. Emeryville, CA: Shoemaker & Hoard.

Bowers, John Z. 1981. "The Odyssey of Smallpox Vaccination." *Bulletin of the History of Medicine* 55: 17–33.

Brandt, Allan M. 1987. *No Magic Bullet: A Social History of Venereal Disease in the United States since 1880*. New York: Oxford University Press.

Braun, Bruce. 2007. "Biopolitics and the Molecularization of Life." *Cultural Geographies* 14: 6–28.

———. 2011. "Governing Disorder: Biopolitics and the Molecularization of Life." In *Global Political Ecology*, edited by Richard Peet, Paul Robbins, and Michael Watts, 389–411. London: Routledge.

Braun, Bruce, and Sarah J. Whatmore. 2010. "The Stuff of Politics: An Introduction." In *Political Matter: Technoscience, Democracy, and Public Life*, edited by Bruce Braun and Sarah J. Whatmore, ix-xxxix. Minneapolis: University of Minnesota Press.

Brier, Jennifer. 2011. *Infectious Ideas: U.S. Political Responses to the AIDS Crisis*. Chapel Hill: University of North Carolina Press.

Buchanan, Allen, and Maureen C. Kelley. 2011. "Biodefence and the Production of Knowledge: Rethinking the Problem." *Journal of Medical Ethics* 39(4): 195–204.

Bull, Malcolm. 2004. "States Don't Really Mind Their Citizens Dying (Provided They Don't All Do It at Once): They Just Don't Like Anyone Else to Kill Them." *London Review of Books*, December 16, 3–6.

Burgan, Mary. 2002. "Contagion and Culture: A View from Victorian Studies." *American Literary History* 14(4): 837–44.

Bush, Larry M., and Maria T. Perez. 2012. "The Anthrax Attacks 10 Years Later." *Annals of Internal Medicine* 156(1): 41–45.

Butler, Declan. 2008. "Politically Correct Names Given to Flu Viruses." *Nature* 452(7190): 923.

Butler, Judith, and Gayatri Chakravorty Spivak. 2007. *Who Sings the Nation-State? Language, Politics, Belonging*. London: Seagull Books.

Campbell, David. 1992. *Writing Security: United States Foreign Policy and the Politics of Identity*. Minneapolis: University of Minnesota Press.

Cañada, Jose A. 2013. "A Bio-objects Approach to Biosecurity: The 'Mutant Flu' Controversy as a Bioobjectification Process." *Croatian Medical Journal* 54: 592–97.

Carey, James. 1977. "Mass Communication Research and Cultural Studies." In *Mass Communication and Society*, edited by James Curran, Michael Gurevitch, and Janet Woollacott, 315–48. London: Edward Arnold.

———. 1981. "McLuhan and Mumford: The Roots of Modern Media Analysis." *Journal of Communication* 31(3): 162–78.

Caton, Hiram. 2000. "Thought Contagion: How Belief Spreads through Society." *Politics and the Life Sciences* 19 (2).

Cello, Jeronimo, Aniko V. Paul, and Eckard Wimmer. 2002. "Chemical Synthesis of Poliovirus CDNA: Generation of Infectious Virus in the Absence of Natural Template." *Science* 297(5583): 1016.

Choffnes, Eileen R., Alison Mack, and David A. Relman. 2010. "The Domestic and International Impacts of the 2009-H1N1 Influenza *A* Pandemic: Global Challenges, Global Solutions." Washington, D.C.: National Academies Press.

Clark, William R. 2008. *Bracing for Armageddon? The Science and Politics of Bioterrorism in America.* Oxford: New York.

Clarke, Richard A., and Rand Beers. 2006. *The Forgotten Homeland: A Century Foundation Task Force Report.* New York: Century Foundation Press.

Cohen, Ed. 2009. *A Body Worth Defending: Immunity, Biopolitics, and the Apotheosis of the Modern Body.* Durham, NC: Duke University Press.

Cole, Leonard A. 2003. *The Anthrax Letters.* Washington, DC: Joseph Henry Press.

Collett, Mark S. 2006. "Impact of Synthetic Genomics on the Threat of Bioterrorism with Viral Agents." In *Working Papers for Synthetic Genomics: Risks and Benefits for Science and Society,* edited by Michele Garfinkel, Drew Endy, Gerald Epstein and Robert Friedman, 83–103. http://hdl.handle.net/1721.1/39658.

Collier, Stephen J. 2008. "Enacting Catastrophe: Preparedness, Insurance, Budgetary Rationalization." *Economy and Society* 37(2): 224–50.

———. 2011. *Post-Soviet Social Neoliberalism, Social Modernity, Biopolitics.* Princeton, NJ: Princeton University Press.

Collier, Stephen J., Andrew Lakoff, and Paul Rabinow. 2004. "Biosecurity: Towards an Anthropology of the Contemporary." *Anthropology Today* 20(5): 3–7.

Comaroff, Jean. 2007. "Beyond Bare Life: AIDS, (Bio)Politics, and the Neoliberal Order." *Public Culture* 19(1): 197–219.

Coole, Diane, and Samantha Frost, eds. 2010. *New Materialisms: Ontology, Agency, and Politics.* Durham, NC: Duke University Press.

Cooper, Melinda. 2006. "Pre-empting Emergence." *Theory, Culture and Society* 23(4): 113–35.

———. 2008. *Life as Surplus: Biotechnology and Capitalism in the Neoliberal Era.* Seattle: University of Washington Press.

Correia, David. 2013. "F**k Jared Diamond." *Capitalism Nature Socialism* 24(4): 1–6.

Covert, Norman. 1993. *Cutting Edge: A History of Fort Detrick, Maryland, 1943–1993.* Fort Detrick, MD: Public Affairs Office, Headquarters U.S. Army Garrison.

Crisp, Alastair, Chiara Boschetti, Malcom Perry, Alan Tunnacliffe, and Gos Micklem. 2015. "Expression of Multiple Horizontally Acquired Genes is a Hallmark of Both Vertebrate and Invertebrate Genomes." *Genome Biology* 16(50): 1–13.

Crosby, Alfred W. 1972. *The Columbian Exchange: Biological and Cultural Consequences of 1492*. Westport, CT: Greenwood Press.

———. 1986. *Ecological Imperialism: The Biological Expansion of Europe, 900–1900*. Cambridge: Cambridge University Press.

Cueto, Marcos. 2007. *Cold War, Deadly Fevers: Malaria Eradication in Mexico, 1955–1975*. Washington, DC: Woodrow Wilson Center Press.

Daase, Christopher, and Oliver Kessler. 2007. "Knowns and Unknowns in the 'War on Terror': Uncertainty and the Political Construction of Danger." *Security Dialogue* 38(4): 411–34.

D'Arcy, Michael B. 2006. *Protecting the Homeland, 2006/2007*. Washington, DC: Brookings Institution Press.

Davis, Cynthia J. 2002. "Contagion as Metaphor." *American Literary History* 14(4): 828–36.

Davis, Mike. 2005. *The Monster at Our Door: The Global Threat of Avian Flu*. New York: New Press.

Davis, Tracy C. 2007. *Stages of Emergency: Cold War Nuclear Civil Defense*. Durham, NC: Duke University Press.

Dean, Mitchell. 2004. *Governmentality: Power and Rule in Modern Society*. London: Sage.

De Goede, Marieke. 2008. "Beyond Risk: Premediation and the Post-9/11 Security Imagination." *Security Dialogue* 39(2–3): 155–76.

Deleuze, Giles, and Felix Guattari. 1988. *A Thousand Plateaus: Capitalism and Schizophrenia*. Translated by B. Massumi. London: Athlone Press.

"Department of Homeland Security Announces TOPOFF 3." 2004. *Journal of Environmental Health* 67(1): 53.

Der Derian, James. 2001. *Virtuous War: Mapping the Military Industrial Media Entertainment Network*. Boulder, CO: Westview Press.

Dick, Marshall, and Everett Hanel Jr. 1970. *Design Criteria for Microbiological Facilities, Fort Detrick, Maryland*, vols. 1 and 2. Department of the Army, Technical Engineering Division Project.

Dillon, Michael. 2003. "Virtual Security: A Life Science of (Dis)Order." *Millennium: Journal of International Studies* 32: 531.

———. 2008. "Underwriting Security." *Security Dialogue* 39(2–3): 309–32.

Diprose, Rosalyn, Niamh Stephenson, Catherine Mills, Kane Race, and Gay Hawkins. 2008. "Governing the Future: The Paradigm of Prudence in Political Technologies of Risk Management." *Security Dialogue* 39(2–3): 267–88.

Dixon, David. 2005. *Never Come to Peace Again: Pontiac's Uprising and the Fate of the British Empire in North America*. Norman: University of Oklahoma Press.

"Don't Underestimate the Enemy." 2001. *Nature* 409(6818): 269.

Doolen, Andy. 2004. "Reading and Writing Terror: The New York Conspiracy Trials of 1741." *American Literary History* 16(3): 377–406.

Dower, Nigel. 2002. "Against War as a Response to Terrorism." *Philosophy and Geography* 5(1): 29–34.

Druett, H. A., D. W. Henderson, L. Packman, and S. Peacock. 1953. "Studies on Respiratory Infection, I: The Influence of Particle Size on Respiratory Infection with Anthrax Spores." *Journal of Hygiene* 51: 359.

Elbe, Stefan. "HIV/AIDS and Security: A Biopolitical Perspective." 2006. *Conference Papers: International Studies Association.* San Diego, CA. March 22.

———. "Risking Lives: AIDS, Security and Three Concepts of Risk." 2008. *Security Dialogue* 39(2–3): 177–98.

Enemark, Christian. *Disease and Security: Natural Plagues and Biological Weapons in East Asia.* London: Routledge, 2007.

Epstein, Gerald L. 2012. "Biosecurity 2011: Not a Year to Change Minds." *Bulletin of the Atomic Scientists* 68(1): 29–38.

Epstein, Joshua M., Derek A.T. Cummings, Shubha Chakravarty, Ramesh M. Singha, and Donald S. Burke. 2002. "Toward a Containment Strategy for Smallpox Bioterror: An Individual-Based Computational Approach." Johns Hopkins University–Brookings Institution, Center on Social and Economic Dynamics, Working Paper No. 31. December. www.brookings.edu/~/media/research/files/reports/2002/12/terrorism/bioterrorism.pdf.

Erickson, Christian W., and Bethany A. Barratt. 2004. "Prudence or Panic? Preparedness Exercises, Counterterror Mobilization, and Media Coverage: Dark Winter, TOPOFF 1 and 2." *Journal of Homeland Security and Emergency Management* 1(4): 1–21.

Ericson, Richard and Kevin D. Haggerty. 2006. *The New Politics of Surveillance and Visibility.* Toronto: University of Toronto Press.

Ewald, François. 1993. "Two Infinities of Risk." In *The Politics of Everyday Fear,* edited by Brian Massumi, 221–28. Minneapolis: University of Minnesota Press.

Fearnley, Lyle. 2008. "Redesigning Syndromic Surveillance for Biosecurity." In *Biosecurity Interventions: Global Health and Security in Question,* edited by Andrew Lakoff and Stephen J. Collier, 61–88. New York: Columbia University Press.

Feldman, Ilana, and Miriam Ticktin, eds. 2010. *In the Name of Humanity: The Government of Threat and Care.* Durham, NC: Duke University Press.

Fenn, Elizabeth A. 2001. *Pox Americana: The Great Smallpox Epidemic of 1775–82.* New York: Hill and Wang.

Fenner, Frank. 1988. *Smallpox and Its Eradication.* Geneva: World Health Organization.

Ferguson, Neal M., Matt J. Keeling, W. John Emunds, Raymond Gani, Bryan T. Grenfell, Roy M. Anderson, and Steve Leach. 2003. "Planning for Smallpox Outbreaks." *Nature* 425(6959): 681–85.

Foege, William H. 2011. *House on Fire: The Fight to Eradicate Smallpox.* Berkeley: University of California Press.

Fortun, Kim. 2001. *Advocacy after Bhopal: Environmentalism, Disaster, New Global Orders.* Chicago: University of Chicago Press, 2001.

Fothergill, Leroy D. 1957. "Biological Warfare and Its Defense." *Public Health Reports* 72: 865.

———. "Biological Warfare: Nature and Consequences." 1964. *Texas State Journal of Medicine,* January, 8–14.

Foucault, Michel. 1972. *The Archaeology of Knowledge.* Translated by Alan Sheridan. New York: Pantheon Books.

———. 1997. *Society Must Be Defended: Lectures at the Collège de France, 1975–1976*. New York: Picador.

———. 2007. *Security, Territory, Population: Lectures at the Collège de France, 1977–78*. Translated by Graham Burchell, edited by Michel Senellart, François Ewald, and Alessandro Fontana. Basingstoke, UK: Palgrave Macmillan.

Franklin, Sarah. 1995. "Science as Culture, Cultures of Science." *Annual Review of Anthropology* 24: 163–84.

Friedberg, Aaron L. 2000. *In the Shadow of the Garrison State: America's Anti-statism and Its Cold War Grand Strategy*. Princeton, NJ: Princeton University Press.

Füredi, Frank. 2002. *Culture of Fear: Risk-Taking and the Morality of Low Expectation*. London: Continuum.

Garrett, Laurie. 1994. *The Coming Plague: Newly Emerging Diseases in a World out of Balance*. New York: Farrar, Straus and Giroux.

Geissler, Erhard, and John Ellis van Courtland Moon, eds. 1999. *Biological and Toxin Weapons: Research, Development and Use from the Middle Ages to 1945*. New York: Oxford University Press.

Geyer, Michael. 1989. "The Militarization of Europe, 1914–1945." In *The Militarization of the Western World*, edited by John Gillis, 75–80. New Brunswick, NJ: Rutgers University Press.

Gibbs, Adrian J., John S. Armstrong, and Jean C. Downie. 2009. "From Where Did the 2009 'Swine Origin' Influenza A Virus (H1N1) Emerge?" *Virology Journal* 6: 207.

Giddens, Anthony. 1999. "Risk and Responsibility." *Modern Law Review* 62(1): 1–10.

Giroux, Henry A. 2008. "Beyond the Biopolitics of Disposability: Rethinking Neoliberalism in the New Gilded Age." *Social Identities* 14(5): 587–620.

Golinski, Jan. 1998. *Making Natural Knowledge: Constructivism and the History of Science*. Cambridge: Cambridge University Press.

Gorman, Siobhan, and Marilyn Werber Serafini. 2003. "Homeland Security's Reality TV Show." *National Journal* 35(19): 1482.

Graham, Bob, James M. Talent, and Graham T. Allison. 2008. *World at Risk: The Report of the Commission on the Prevention of WMD Proliferation and Terrorism*. New York: Vintage Books.

Gray, Colin S. 1994. "Strategy in the Nuclear Age: The United States, 1945–1991." In *The Making of Strategy: Rulers, States, and War*, edited by Williamson Murray, MacGregor Knox, and Alvin H. Bernstein, 579–613. Cambridge: Cambridge University Press.

Green, Stephen. 1999. "A Plague on the Panopticon: Surveillance and Power in the Global Information Economy." *Information, Communication and Society* 2(1): 26–44.

Gregory, Steven. 2007. *The Devil behind the Mirror: Globalization and Politics in the Dominican Republic*. Berkeley: University of California Press.

Grenier, John. 2005. *The First Way of War: American War Making on the Frontier, 1607–1814*. Cambridge: Cambridge University Press.

Grossman, Andrew D. 2001. *Neither Dead nor Red: Civilian Defense and American Political Development During the Early Cold War*. New York: Routledge.

Guillemin, Jeanne. 2005. "Inventing Bioterrorism: The Political Construction of Civilian Risk." In *Making Threats: Biofears and Environmental Anxieties*, edited by Betsy Hartmann, Banu Subramaniam, and Charles Zerner. Lanham, MD: Rowman and Littlefield.

Gupta, Akhil, and James Ferguson, eds. 1997. *Anthropological Locations: Boundaries and Grounds of a Field Science*. Berkeley: University of California Press.

Gusterson, Hugh. 2008. "Nuclear Futures: Anticipatory Knowledge, Expert Judgment, and the Lack That Cannot Be Filled." *Science and Public Policy* 35(8): 551–60.

Gutierrez, Christopher M. 2006. "Bodies of Terror/Terrorizing Bodies." MA thesis, Concordia University.

Habermas, Jürgen, Jacques Derrida, and Giovanna Borradori. 2003. *Philosophy in a Time of Terror: Dialogues with Jürgen Habermas and Jacques Derrida*. Chicago: University of Chicago Press.

Hacking, Ian. "Risk and Dirt." 2003. In *Risk and Morality*, edited by Richard Ericson and Aaron Doyle, 22–47. Toronto: University of Toronto Press.

Haggerty, Kevin D. 2006. "Tear Down the Walls: On Demolishing the Panopticon." In *Theorizing Surveillance: The Panopticon and Beyond*, edited by David Lyon, 23–45. Cullompton, UK: Willan Publishing.

Haggerty, Kevin D., and Richard V. Ericson. 2000. "The Surveillant Assemblage." *British Journal of Sociology* 51(4): 605–22.

Hall, Stuart. 1977. "Culture, the Media and the 'Ideological Effect.'" In *Mass Communication and Society*, edited by James Curran, Michael Gurevitch, and Janet Woollacott, 315–48. London: Edward Arnold.

Haraway, Donna. 1991. *Simians, Cyborgs, and Women: The Reinvention of Nature*. New York: Routledge.

Hård, Mikael, and Andrew Jamison. 2005. *Hubris and Hybrids: A Cultural History of Technology and Science*. New York: Routledge.

Harper, G.J., and J.D. Morton. 1953. "The Respiratory Retention of Bacterial Aerosols: Experiments with Radioactive Spores." *Journal of Hygiene* 51(3): 372–85.

Hartmann, Betsy, Banu Subramaniam, and Charles Zerner, eds. 2005. *Making Threats: Biofears and Environmental Anxieties*. Lanham, MD: Rowman and Littlefield.

Harvey, David. 1990. *The Condition of Postmodernity*. Cambridge: Blackwell.

Heath, Deborah, Rayna Rapp, and Karen-Sue Taussig. 2004. "Genetic Citizenship." In *A Companion to the Anthropology of Politics*, edited by David Nugent and Joan Vincent, 152–67. Malden, MA: Blackwell.

Heaton, John A., Anne M. Murphy, Susan Allan, and Harald Pietz. 2003. "Legal Preparedness for Public Health Emergencies: TOPOFF 2 and Other Lessons." *Journal of Law, Medicine and Ethics* 31(4): 43–44.

Helmreich, Stefan. 2005. "Biosecurity: A Response to Collier, Lakoff and Rabinow." *Anthropology Today* 21(2): 21.

Henderson, Donald A. 2009. *Smallpox: The Death of a Disease: The Inside Story of Eradicating a Worldwide Killer*. Amherst, NY: Prometheus Books.

Hess, David J., Linda L. Layne, and Arie Rip. 1992. *Knowledge and Society: The Anthropology of Science and Technology*. London: AJI Press.

Hitchcock, Ethan Allen. 1909. *Fifty Years in Camp and Field: Diary of Major-General Ethan Allen Hitchcock*. Edited by W.A. Croffut. New York: G.P. Putnam's Sons.

Hopkins, Donald R. 1983. *Princes and Peasants: Smallpox in History*. Chicago: University of Chicago Press.

Horkheimer, Max, and Theodor Adorno. 2007. *Dialectic of Enlightenment: Cultural Memory in the Present*. Edited by Gunzelin Schmid Noerr, translated by Edmund Jephcott. Stanford, CA: Stanford University Press.

Hounshell, David A. 2001. "Epilogue: Rethinking the Cold War; Rethinking Science and Technology in the Cold War; Rethinking the Social Study of Science and Technology." *Social Studies of Science* 31(2): 289–97.

Humphreys, Margaret. 2002. "No Safe Place: Disease and Panic in American History." *American Literary History* 14(4): 845–57.

Ilchmann, Kai, and James Revill. 2014. "Chemical and Biological Weapons in the 'New Wars.'" *Science and Engineering Ethics* 20(3): 753–67.

Inglesby, Thomas V., Rita Grossman, and Tara O'Toole. 2001. "A Plague on Your City: Observations from TOPOFF." *Clinical Infectious Diseases* 32(3): 435–45.

Jabri, Vivienne. 2006. "War, Security and the Liberal State." *Security Dialogue* 37(1): 47–64.

Jackson, Ronald J., Alistair J. Ramsay, Carina D. Christensen, Sandra Beaton, Diana F. Hall, and Ian A. Ramshaw. 2001. "Expression of Mouse Interleukin-4 by a Recombinant Ectromelia Virus Suppresses Cytolytic Lymphocyte Responses and Overcomes Genetic Resistance to Mousepox." *Journal of Virology* 75(3): 1205–10.

Jarvis, Lee. 2009. "The Spaces and Faces of Critical Terrorism Studies." *Security Dialogue* 40(1): 5–27.

Jasanoff, Sheila. 2010. "Beyond Calculation: A Democratic Response to Risk." In *Disaster and the Politics of Intervention*, edited by Andrew Lakoff, 14–40. New York: Columbia University Press.

Jefferson, Catherine, Filippa Lentzos, and Claire Marris. 2014. "Synthetic Biology and Biosecurity: Challenging the 'Myths.'" *Frontiers in Public Health* 2(115): 1–15.

Jemski, Joseph V., and G. Briggs Phillips. 1963. "Microbiological Safety Equipment." *Laboratory Animal Care* 13: 2–12.

Jenner, Edward. 1798. *An Inquiry into the Causes and Effects of the Variolae Vaccinae, a Disease Discovered in Some of the Western Counties of England, Particularly Gloucestershire, and Known by the Name of the Cow Pox*. London: Printed, for the author, by Sampson Low, and sold by Law [etc.].

Johansson, R.R., and D.H. Ferris. 1946. "Photograph of Airborne Particles during Bacteriological Plating Operations." *New England Journal of Infectious Disease* 78: 238–52.

Kaltman, Stacey, Rochelle E. Tractenberg, Kathryn Taylor, and Bonnie L. Green. 2006. "Modeling Dimensions of Choice in Accepting the Smallpox Vaccine." *American Journal of Health Behavior* 30(5): 513–24.

Kellman, Barry. 2007. *Bioviolence: Preventing Biological Terror and Crime.* New York: Cambridge University Press.

King, Nicholas B. 2002. "Security, Disease, Commerce: Ideologies of Postcolonial Global Health." *Social Studies of Science* 32(5–6): 763.

———. 2003. "The Influence of Anxiety: September 11, Bioterrorism, and American Public Health." *Journal of the History of Medicine* 58: 433–41.

———. 2004."The Scale Politics of Emerging Diseases." *Osiris* 19(1): 62–76.

———. 2005. "The Ethics of Biodefense." *Bioethics* 19(4): 432–46.

Kittelsen, Sonja. 2009. "Conceptualizing Biorisk: Dread Risk and the Threat of Bioterrorism in Europe." *Security Dialogue* 40(1): 51–71.

Koh, Howard K., Loris J. Elqura, Christine M. Judge, John P. Jacob, Amy E. Williams, M. Suzanne Crowther, Richard A. Serino, and John M. Auerbach. 2008. "Implementing the Cities Readiness Initiative: Lessons Learned from Boston." *Disaster Medicine and Public Health Preparedness* 2(1): 40–49.

Koplow, David A. 2003. *Smallpox: The Fight to Eradicate a Global Scourge.* Berkeley: University of California Press.

Kosek, Jake. 2006. *Understories: The Political Life of Forests in Northern New Mexico.* Durham, NC: Duke University Press.

Kraut, Alan M. 1994. *Silent Travelers: Germs, Genes, and the "Immigrant Menace."* Baltimore: Johns Hopkins University Press.

Krisberg, Kim. 2003. "Public Health Preparedness Drills Reap Benefits, Concerns." *Nation's Health* 33(9): 9.

Kuletz, Valerie L. 1998. *The Tainted Desert: Environmental and Social Ruin in the American West.* New York: Routledge.

Kupperman, Robert H. 1989. *Final Warning: Averting Disaster in the New Age of Terrorism.* New York: Doubleday.

Lakoff, Andrew. 2007. "Preparing for the Next Emergency." *Public Culture* 19(2): 247–71.

———. 2008a. "The Generic Biothreat, or, How We Became Unprepared." *Cultural Anthropology* 23(3): 399–428.

———. 2008b. "From Population to Vital System: National Security and the Changing Object of Public Health." In *Biosecurity Interventions: Global Health and Security in Question,* edited by Andrew Lakoff and Stephen J. Collier, 33–60. New York: Columbia University Press.

———. 2010. *Disaster and the Politics of Intervention.* New York: Columbia University Press.

———. 2012. "Epidemic Intelligence: Toward a Genealogy of Global Health Security." In *Contagion: Health, Fear, Sovereignty,* edited by Bruce Magnusson and Zahi Zalloua, 44–70. Seattle: University of Washington Press.

Lakoff, Andrew, and Stephen J. Collier, eds. 2008. *Biosecurity Interventions: Global Health and Security in Question.* New York: Columbia University Press.

———. 2010. "Infrastructure and Event." In *Political Matter: Technosicence, Democracy, and Public Life,* edited by Bruce Braun and Sarah J. Whatmore, 243–66. Minneapolis: University of Minnesota Press.

Larkin, Howard. 2003. "Three Days in May." *Hospitals and Health Networks* 77(7): 24.

Latour, Bruno. 1993. *We Have Never Been Modern*. Cambridge, MA: Harvard University Press.
———. 1999. *Pandora's Hope: Essays on the Reality of Science Studies*. Cambridge, MA: Harvard University Press.
Latour, Bruno, and Steve Woolgar. 1986. *Laboratory Life: The Construction of Scientific Facts*. Princeton, NJ: Princeton University Press.
Lederberg, Joshua. 1999. "Keeping America Secure for the 21st Century." *Procedures of the National Academy of Sciences of the United States of America* 96(7): 3486–88.
Leiss, William. 2007. "Modern Science, Enlightenment, and the Domination of Nature: No Exit?" *Fast Capitalism* 2(2). www.uta.edu/huma/agger/fastcapitalism/2_2/leiss.html, accessed June 27, 2014.
Leitenberg, Milton. 2002. "Biological Weapons and Bioterrorism in the First Years of the Twenty-First Century." *Politics and the Life Sciences* 21(2): 3–27.
Lentzos, Filippa. 2015. "Synthetic Biology's Defense Dollars Signals and Perceptions." *Public Library of Science Blog*. http://blogs.plos.org/synbio/2015/12/24/synthetic-biologys-defence-dollars-signals-and-perceptions/. Accessed February 28, 2016.
Lentzos, Filippa, Catherine Jefferson, and Claire Marris. 2014. "The Myths (and Realities) of Synthetic Bioweapons." *Bulletin of the Atomic Scientists* (September): 1–6.
Lerner, Barron H., and David J. Rothman. 1995. "Medicine and the Holocaust: Learning More of the Lessons." *Annals of Internal Medicine* 122(10): 793–94.
Lewontin, Richard C. 1991. *Biology as Ideology: The Doctrine of DNA*. New York: Harper Collins.
Luhmann, Niklas. 1993. *Risk: A Sociological Theory*. New York: Aldine de Gruyter.
Lynch, Michael. 1995. "Laboratory Space and the Technological Complex: An Investigation of Topical Contextures." In *Ecologies of Knowledge: Work and Politics in Science and Technology*, edited by Susan Leigh Star, 226–56. Albany: State University of New York Press.
Lyon, David. 2003. *Surveillance after September 11*. Malden, MA: Polity Press.
———. 2004. "Technology vs. 'Terrorism': Circuits of City Surveillance since September 11, 2001." In *Cities, War, and Terrorism: Towards an Urban Geopolitics*, edited by Stephen Graham, 296–311. Malden, MA: Blackwell Publishing.
———. 2006. *Theorizing Surveillance: The Panopticon and Beyond*. Cullompton, UK: Willan Publishing.
Maclean, Sandra J. 2008. "Microbes, Mad Cows and Militaries: Exploring the Links between Health and Security." *Security Dialogue* 39(5): 475–94.
Maier, Charles S. 2000. "Consigning the Twentieth Century to History: Alternative Narratives for the Modern Era." *American Historical Review* 105(3): 807–31.
Marcus, George E. 1998. *Ethnography through Thick and Thin*. Princeton, NJ: Princeton University Press.

Marcus, George E., and Michael M. J. Fischer. 1986. *Anthropology as Cultural Critique: An Experimental Moment in the Human Sciences*. Chicago: University of Chicago Press.

Margulis, Lynn. 1998. *Symbiotic Planet: A New Look at Evolution*. New York: Basic Books.

Martin, Emily. 1994. *Flexible Bodies: Tracking Immunity in American Culture from the Days of Polio to the Age of AIDS*. Boston: Beacon Press.

Marx, Karl. 1973. *Capital, Volume 1: A Critique of Political Economy*. Edited by Friedrich Engels, translated by Samuel Moore and Edward Aveling. Moscow: Progress Publishers.

———. 1978. "Estranged Labour." In *The Marx-Engels Reader*, edited by Robert Tucker, 70–81. New York: Norton.

Marx, Karl, and Friedrich Engels. 1975. *Marx-Engels Collected Works, Volume 40*. Moscow: Progress Publishers.

Maryland State Archives. n.d. Nomination of One-Million-Liter Test Sphere (Horton Test Sphere) at Fort Detrick, Frederick, Maryland, for inclusion in the National Register of Historic Places. http://msa.maryland.gov/megafile/msa/stagsere/se1/se5/010000/010400/010489/pdf/msa_se5_10489.pdf.

Masco, Joseph. 2006. *Nuclear Borderlands: The Manhattan Project in Post–Cold War New Mexico*. Princeton, NJ: Princeton University Press.

———. 2008. "'Survival Is Your Business': Engineering Ruins and Affect in Nuclear America." *Cultural Anthropology* 23(2): 361–98.

———. 2010a. "Atomic Health, or How the Bomb Altered American Notions of Death." In *Against Health: How Health Became the New Morality*, edited by Jonathan M. Metzl and Anna Kirkland, 133–53. New York: New York University Press.

———. 2010b. "'Sensitive but Unclassified': Secrecy and the Counterterrorist State." *Public Culture* 22(3): 433–63.

———. 2011. "Mutant Ecologies: Radioactive Life in Post–Cold War New Mexico." In *Global Political Ecology*, edited by Richard Peet, Paul Robbins, and Michael Watts, 285–304. New York: Routledge.

———. 2014. *The Theater of Operations: National Security Affect from the Cold War to the War on Terror*. Durham, NC: Duke University Press.

Massumi, Brian. 2007. "Potential Politics and the Primacy of Preemption." *Theory and Event* 10(2): 1–19.

Masterson, Lori, Christel Steffen, Michael Brin, Mary Frances Kordick, and Steve Christos. 2009. "Willingness to Respond: Of Emergency Department Personnel and Their Predicted Participation in Mass Casualty Terrorist Events." *Journal of Emergency Medicine* 36(1): 43–49.

McAlister, Melani. 2001. *Epic Encounters: Culture, Media, and U.S. Interests in the Middle East since 1945*. Berkeley: University of California Press.

McCarthy, Anna. 2001. *Ambient Television: Visual Culture and Public Space*. Durham, NC: Duke University Press.

McKenzie, F. Ellis. 2004. "Smallpox Models as Policy Tools." *Emerging Infectious Diseases* 10(11): 2044–47.

McLean, Stuart. 2004. *The Event and Its Terrors: Ireland, Famine, Modernity*. Stanford, CA: Stanford University Press.

McNeill, William H. 1976. *Plagues and Peoples*. Garden City, NY: Anchor Press.
Merck, George M. 1946. "Biological Warfare." *The Military Surgeon* 98: 237–42.
Metzl, Jonathan, and Anna Rutherford Kirkland. 2010. *Against Health: How Health Became the New Morality*. New York: New York University Press.
Miller, Judith, Stephen Engelberg, and William Broad. 2001. *Germs: The Ultimate Weapon*. New York: Simon & Schuster.
Miller, Toby. 2007. *Cultural Citizenship: Cosmopolitanism, Consumerism, and Television in a Neoliberal Age*. Philadelphia: Temple University Press.
Mills, Catherine. "Agamben's Messianic Biopolitics: Biopolitics, Abandonment and Happy Life." *Contretemps* 5: 42–62.
Mirzoeff, Nicholas. 2004. *Watching Babylon: The War in Iraq and Global Visual Culture*. New York: Routledge, 2005.
Mitka, Mike. "Bioterror Exercise Tests Agencies' Mettle." *Journal of the American Medical Association* 289(22): 2927.
Molina, Natalia. 2006. *Fit to Be Citizens? Public Health and Race in Los Angeles, 1879–1939*. Berkeley: University of California Press, 2006.
Moore, Donald S. *Suffering for Territory: Race, Place, and Power in Zimbabwe*. Durham, NC: Duke University Press, 2005.
Muller, Benjamin J. 2008. "Securing the Political Imagination: Popular Culture, the Security Dispositif and the Biometric State." *Security Dialogue* 39(2–3): 199–220.
Muntean, Nick. 2009. "Viral Terrorism and Terrifying Viruses: The Homological Construction of the 'War on Terror' and the Avian Flu Pandemic." *International Journal of Media and Cultural Politics* 5(3): 199–216.
Murray, Stuart J. 2006. "Thanatopolitics: On the Use of Death for Mobilizing Political Life." *Polygraph* 18: 191–215.
Myers, Natasha. 2009. "Performing the Protein Fold." In *Simulation and Its Discontents*, edited by Sherry Turkle, 171–88. Cambridge, MA: MIT Press.
Nester, William R. 2000. *"Haughty Conquerors": Amherst and the Great Indian Uprising of 1763*. Westport, CT: Praeger.
Nguyen, Vinh-kim. 2005. "Antiretroviral Globalism, Biopolitics, and Therapeutic Citizenship." In *Global Assemblages: Technology, Politics, and Ethics as Anthropological Problems*, edited by Aihwa Ong and Stephen J. Collier, 124–44. Malden, MA: Blackwell.
Oakes, Guy. 1994. *The Imaginary War: Civil Defense and American Cold War Culture*. New York: Oxford University Press.
O'Day, Alan, ed. 2004. *Weapons of Mass Destruction and Terror*. Burlington, VT: Ashgate.
Ojakangas, Mika. 2005. "Impossible Dialogue on Bio-power: Agamben and Foucault." *Foucault Studies* 2: 5–28.
Olson, Debra, Aggie Leitheiser, Christopher Atchison, Susan Larson, and Cassandra Homzik. 2005. "Public Health and Terrorism Preparedness: Cross-Border Issues." *Public Health Reports* 120: 76–83.
Ong, Aihwa. 2006. *Neoliberalism as Exception: Mutations in Citizenship and Sovereignty*. Durham, NC: Duke University Press.

Ong, Aihwa, and Stephen J. Collier. 2005. *Global Assemblages: Technology, Politics, and Ethics as Anthropological Problems.* Malden, MA: Blackwell.

O'Toole, Tara, Michael Mair, and Thomas Inglesby. 2002. "Shining Light on 'Dark Winter.'" *Clinical Infectious Diseases* 34: 972–83.

Parisi, Luciana. 2007. "Biotech: Life by Contagion." *Theory, Culture and Society* 24(6): 29–52.

Parr, Hester. 2002. "New Body-Geographies: The Embodied Spaces of Health and Medical Information on the Internet." *Environment and Planning: Society and Space* 20: 73–95.

Peach, James, and Kenneth Nowotny. 1982. "Economic Development in Border Areas: The Case of New Mexico and Chihuahua." *Journal of Economic Issues* 16(2): 489–96.

Pearson, Helen. 2003. "Hollywood Horrors Bring Bioterror to Life." *Nature Medicine* 9(5): 489.

Peckham, Howard H. 1947. *Pontiac and the Indian Uprising.* Princeton, NJ: Princeton University Press.

Peet, Richard, Paul Robbins, and Michael Watts, eds. 2011. *Global Political Ecology.* New York: Routledge.

Pernick, Martin S. 2002. "Contagion and Culture." *American Literary History* 14(4): 858–65.

Peters, Stephanie True. 2005. *Smallpox in the New World.* New York: Benchmark Books.

Petryna, Adriana. 2002. *Life Exposed: Biological Citizens after Chernobyl.* Princeton, NJ: Princeton University Press.

Philip, Robert N. 2000. *Rocky Mountain Spotted Fever in Western Montana: Anatomy of a Pestilence.* Stevensville, MT: Stoneydale Publishing.

Phillips, G. Briggs, Robert W. Edwards, Martin S. Favero, Robert K. Hoffman, Tom B. Lanahan, Norman H. MacLeod, Joseph J. McDade, and Peter Skaliy. 1965. "Microbiological Contamination Control: A State of the Art Report." *Journal of the American Association for Contamination Control* 4: 16–25.

Pike, R.M., and S.E. Sulkin. 1952. "Occupational Hazards in Microbiology." *Science Monthly* 75: 222–28.

Preston, Richard. 1994. *The Hot Zone.* New York: Random House.

———. 2000. *The Demon in the Freezer.* New York: Doubleday.

Price, Don K. 1965. *The Scientific Estate.* Cambridge, MA: Harvard University Press.

Prigogine, Ilya. 1999. "Science in a World of Limited Predictability." In *Ecology: Key Concepts in Critical Theory,* edited by Carolyn Merchant, 401–6. Amherst, NY: Humanity Books.

Rabinow, Paul. 1986. "Representations Are Social Facts." In *Writing Culture: The Poetics and Politics of Ethnography,* edited by James Clifford and George E. Marcus, 234–61. Berkeley: University of California Press.

———. 1996. *Essays on the Anthropology of Reason.* Princeton, NJ: Princeton University Press.

———. 2005. "Midst Anthropology's Problems." In *Global Assemblages: Technology, Politics and Ethics as Anthropological Problems*, edited by Aihwa Ong and Stephen J. Collier, 40–53. Malden, MA: Blackwell,

Rabinow, Paul, and Talia Dan-Cohen. 2005. *A Machine to Make a Future: Biotech Chronicles*. Princeton, NJ: Princeton University Press.

Rajan, Kaushik Sunder. 2006. *Biocapital: The Constitution of Postgenomic Life*. Durham, NC: Duke University Press.

Raney, David. 2003. "'No Ties except Those of Blood': Class, Race, and Jack London's American Plague." In *Papers on Language and Literature* 39(4): 390–430.

Ranlet, Philip. 2000. "The British, the Indians, and Smallpox: What Actually Happened at Fort Pitt in 1763?" *Pennsylvania History* 67(3): 427–41.

Ratner, Daniel, and Mark A. Ratner. 2004. *Nanotechnology and Homeland Security: New Weapons for New Wars*. Upper Saddle River, NJ: Prentice Hall/PTR.

Razzell, Peter E. 1977. *Edward Jenner's Cowpox Vaccine: The History of a Medical Myth*. Firle, UK: Caliban Books.

Reid, Julian. 2006. "Life Struggles: War, Discipline, and Biopolitics in the Thought of Michel Foucault." *Social Text* 24(1): 129–30.

Reitman, Morton. 1956. "Laboratory Hazards." *American Journal of Medical Technology* 22: 12–13.

Reitman, Morton, and Arnold G. Wedum. 1956. "Microbiological Safety." *Public Health Reports* 71: 659–65.

Rhodes, Lorna A. 1998. "Panoptical Intimacies." *Public Culture* 10(2): 285–311.

Robbins, Paul. 2003. "Networks and Knowledge Systems: An Alternative to 'Race or Place.'" *Antipode* 35(4): 818–22.

Robin, Corey. 2004. *Fear: The History of a Political Idea*. Oxford: Oxford University Press.

Roman, Peter J. 2002. "The Dark Winter of Biological Terrorism." *Orbis* 46(3): 469–82.

Rose, Nikolas. 2001. "The Politics of Life Itself." *Theory, Culture and Society* 18(6): 1–30.

———. 2007. *The Politics of Life Itself: Biomedicine, Power, and Subjectivity in the Twenty-First Century*. Princeton, NJ: Princeton University Press.

Rose, Nikolas, and Carlos Novas. 2005. "Biological Citizenship." In *Global Assemblages: Technology, Politics, and Ethics as Anthropological Problems*, edited by Aihwa Ong and Stephen Collier, 439–63. Malden, MA: Blackwell.

Rozsa, Lajos. 2000. "Spite, Xenophobia, and Collaboration between Hosts and Parasites." *Oikos* 91(2): 396–400.

———. 2009. "The Motivation for Biological Aggression Is an Inherent and Common Aspect of the Human Behavioural Repertoire." *Medical Hypotheses* 72: 217–19.

———. 2014. "A Proposal for the Classification of Biological Weapons Sensu Lato." *Theory in Biosciences* 133(3–4): 129–34.

Rudman, Warren B., Gary Hart, and Stephen E. Flynn. 2002. *America: Still Unprepared, Still in Danger*. New York: Council on Foreign Relations.

Safina, Carl. 1997. *Songs for the Blue Ocean: Encounters Along the World's Coasts and Beneath the Seas*. New York: Holt Paperbacks.

Samimian-Darash, Limor. 2009. "A Pre-event Configuration for Biological Threats: Preparedness and the Constitution of Biosecurity Events." *American Ethnologist* 36(3): 478–91.

Sarasin, Philipp. 2006. *Anthrax: Bioterror as Fact and Fantasy*. Cambridge, MA: Harvard University Press.

Savage, Rowan. 2007. "Disease Incarnate": Biopolitical Discourse and Genocidal Dehumanisation in the Age of Modernity." *Journal of Historical Sociology* 20(3): 404–40.

Sayres, Sohnya. 2002. "Science under Glass." *American Literary History* 14(1): 160–80.

Scarry, Elaine. 1985. *The Body in Pain*. New York: Oxford.

Schell, Heather. 2002. "The Sexist Gene: Science Fiction and the Germ Theory of History." *American Literary History* 14(4): 805–27.

Schneier, Bruce. 2003. *Beyond Fear: Thinking Sensibly about Security in an Uncertain World*. New York: Copernicus Books.

Schuler, Ari. 2004. "Billions for Biodefense: Federal Agency Biodefense Funding, FY2001–FY2005." *Biosecurity and Bioterrorism* 2(2): 86–96.

Shah, Nayan. 2001. *Contagious Divides: Epidemics and Race in San Francisco's Chinatown*. Berkeley: University of California Press.

Shapiro, Michael J. 2007. "The New Violent Cartography." *Security Dialogue* 38(3): 291–313.

Shelton, Shoshana R., Kathryn Connor, Lori Uscher-Pines, Francesca Matthews Pillemer, James M. Mullikin, and Arthur L. Kellermann. 2012. "Bioterrorism and Biological Threats Dominate Federal Health Security Research." *Health Affairs* 321(12): 2755–63.

Simon, Jeffrey D. 2001. *The Terrorist Trap: America's Experience with Terrorism*. Bloomington: Indiana University Press.

Simpson, David M. 2002. "Earthquake Drills and Simulations in Community-based Training and Preparedness Programmes." *Disasters* 26(1): 55–69.

Sluyter, Andrew. 2003. "NeoEnvironmental Determinism, Intellectual Damage Control, and Nature/Society Science." *Antipode* 35(4): 813–17.

Smith, Michael M. 1974. "The 'Real Expedicion Maritima de la Vacuna' in New Spain and Guatemala." *Transactions of the American Philosophical Society* 64(1): 1–74.

Soja, Edward W. 1989. *Postmodern Geographies: The Reassertion of Space in Critical Social Theory*. London: Verso.

Solovey, Mark. 2001. "Introduction: Science and the State during the Cold War: Blurred Boundaries and a Contested Legacy." *Social Studies of Science* 31(2): 165–70.

Staiger, Janet. 2005. *Media Reception Studies*. New York: New York University Press.

Stern, Alexandra Minna. 1999. "Buildings, Boundaries, and Blood: Medicalization and Nation-Building on the U.S.-Mexico Border, 1910–1930." *Hispanic American Historical Review* 79(1): 41–81.

Stern, Jessica. 2003. "Dreaded Risks and the Control of Biological Weapons." *International Security* 27(3): 89–123.

Stoler, Ann Laura. 2001. "Tense and Tender Ties: The Politics of Comparison in North American History and (Post)Colonial Studies." *Journal of American History* 88(3): 829–65.

———. 2006. "On Degrees of Imperial Sovereignty." *Public Culture* 18(1): 147–84.

Stoneman, Scott. 2007. "Pedagogy in a Time of Terror: Henry Giroux's *Beyond the Spectacle of Terrorism*." *Review of Education, Pedagogy and Cultural Studies* 29(1): 111–35.

Sturken, Marita, Douglas Thomas, and Sandra J. Ball-Rokeach, eds. 2004. *Technological Visions: The Hopes and Fears That Shape New Technologies.* Philadelphia: Temple University Press.

Tambornino, John. 1999. "Locating the Body: Corporeality and Politics in Hannah Arendt." *Journal of Political Philosophy* 7(2): 172.

Tanner, Laura E. 2002. "Bodies in Waiting: Representations of Medical Waiting Rooms in Contemporary American Fiction." *American Literary History* 14(1): 115–30.

Taylor, Nick. 2003. *Review of the Use of Models in Informing Disease Control Policy Development and Adjustment.* Reading, UK: School of Agriculture, Policy and Development, University of Reading. www.veeru.rdg.ac.uk /documents/UseofModelsinDiseaseControlPolicy.pdf.

Teclaw, Robert F. 1979. "Epidemic Modeling." In *A Study of the Potential Economic Impact of Foot and Mouth Disease in the United States,* edited by Earl Hunt McCauley, John C. New Jr., Nasser A. Aulaqi, W.B. Sundquist, and William M. Millers, 103–11. St. Paul: University of Minnesota Press. http://naldc.nal.usda.gov/download/CAT87201575/PDF.

Thornton, William H. 2003. "Cold War II: Islamic Terrorism as Power Politics." *Antipode* 35(2): 205–11.

Tomes, Nancy. 1999. *The Gospel of Germs: Men, Women, and the Microbe in American Life.* Cambridge, MA: Harvard University Press.

———. 2002. "Epidemic Entertainment: Disease and Popular Culture in Early-Twentieth-Century America." *American Literary History* 14(4): 625–52.

Trust for America's Health. 2007. *Ready or Not? Protecting the Public's Health from Diseases, Disasters, and Bioterrorism.* Washington, DC: *Trust for America's Health.* http://healthyamericans.org/report/101/.

Tsing, Anna. 2005. *Friction: An Ethnography of Global Connection.* Princeton, NJ: Princeton University Press.

Tucker, Jonathan B. 2002. *Scourge: The Once and Future Threat of Smallpox.* New York: Grove Press.

———. 2003. "Biosecurity: Limiting Terrorist Access to Deadly Pathogens." Washington, DC: United States Institute of Peace.

Tucknott, Darr. 2006. "Rocky Mountain Laboratories: An Inquiry into Community Opposition to a Biosafety Level IV Expansion." MA Thesis, University of Montana.

Turkle, Sherry, ed. *Simulation and Its Discontents*. Cambridge, MA: MIT Press, 2009.

U.S. Congress. House. Subcommittee on National Security, Veterans Affairs, and International Relations—Committee on Government Reform. 1999. *Combating Terrorism: Assessing the Threat*. 106th Congress. October 20.

U.S. Congress. House. Subcommittee on Prevention of Nuclear and Biological Attack of the Committee on Homeland Security. 2005. *Engineering Bioterror Agents: Lessons from the Offensive U.S. and Russian Biological Weapons Programs*, 109th Congress. July 13.

U.S. Department of Health, Education, and Welfare. n.d. *Biohazard Control and Containment in Oncogenic Virus Research*. Brochure. Center for the History of Microbiology/American Society for Microbiology archives.

U.S. Department of Homeland Security. 2004. *Securing Our Homeland: U.S. Department of Homeland Security Strategic Plan*. Washington: GPO.

U.S. Executive Office of the President, Office of Science and Technology Policy. 2005. *Science and Technology: A Foundation for Homeland Security*. Washington, DC.

U.S. National Research Council. 2002. *Making the Nation Safer: The Role of Science and Technology in Countering Terrorism*. Washington, DC: National Academy Press.

U.S. National Research Council, and Institute of Medicine. 2006. *Globalization, Biosecurity, and the Future of the Life Sciences*. Washington, DC: National Academies Press.

Utley, Robert Marshall. 1984. *The Indian Frontier of the American West, 1846–1890*. Albuquerque: University of New Mexico Press.

Vogel, Kathleen. 2006. "Bioweapons Proliferation: Where Science Studies and Policy Collide." *Social Studies of Science* 36(5): 659–90.

Vogel, Sydney. 2008. "War Medicine: Spain, 1936–1939." *American Journal of Public Health* 98: 2146–49.

Voyles, Traci Brynne. 2015. *Wastelanding: Legacies of Uranium Mining in Navajo Country*. Minneapolis: University of Minnesota Press.

Wade, Beth. 2000. "Drill Tests Response to Terrorism." *American City and County* 115(11): 46.

Wald, Priscilla. 2002a. "Communicable Americanism: Contagion, Geographic Fictions, and the Sociological Legacy of Robert E. Park." *American Literary History* 14(4): 653–85.

———. 2002b. "Introduction: Contagion and Culture." *American Literary History* 14(4): 617–24.

———. 2008. *Contagious: Cultures, Carriers, and the Outbreak Narrative*. Durham, NC: Duke University Press.

———. 2012. "Bio Terror: Hybridity in the Horror Narrative, or What We Can Learn from Monsters." In *Contagion: Health, Fear, Sovereignty*, edited by Bruce Magnusson and Zahi Zalloua, 99–122. Seattle: University of Washington Press.

Waldby, Cathy. 2000. *The Visible Human Project: Informatic Bodies and Posthuman Medicine*. New York: Routledge.

Walters, Ronald G., ed. 1997. *Scientific Authority and Twentieth-Century America:* Baltimore: Johns Hopkins University Press.
Washington, George, Philander D. Chase, Dorothy Twohig, Frank E. Grizzard, and Edward G. Lengel. 1985. *The Papers of George Washington.* Revolutionary War Series. Charlottesville: University Press of Virginia.
Weart, Spencer R. 1988. *Nuclear Fear: A History of Images.* Cambridge, MA: Harvard University Press.
Wedum, Arnold G. 1953. "Bacteriological Safety." *American Journal of Public Health* 43(11): 1428–37.
———. 1959. "Policy, Responsibility and Practice in Laboratory Safety." In *Proceedings of the Second Symposium on Gnotobiotic Technology*, 105–19. Notre Dame, IN: University of Notre Dame Press.
Wedum, Arnold G., and G. Briggs Phillips. 1964. "Criteria for Design of a Microbiological Research Laboratory." *Journal of American Society of Heating, Refrigeration, and Air Conditioning* 6: 46–52.
Wedum, Arnold G., Everett Hanel, G. Briggs Phillips, and Orrin T. Miller. 1956. "Laboratory Design for Study of Infectious Disease." *American Journal of Public Health* 46: 1102–13.
Weindling, Paul. 2000. *Epidemics and Genocide in Eastern Europe, 1890–1945.* New York: Oxford University Press.
Wheelis, Mark, Lajos Rozsa, and Malcolm Dando, eds. 2006. *Deadly Cultures: Biological Weapons since 1945.* Cambridge, MA: Harvard University Press.
White, Richard. 1995. *The Organic Machine.* New York: Hill and Wang.
White, Robert A. 1983. "Mass Communication and Culture: Transition to a New Paradigm." *Journal of Communication* 33(3): 279–301.
Wiant, Chris J. 2000. "Operation TOPOFF: Lessons on Responding to Bioterrorism." *Journal of Environmental Health* 63(3): 50.
Williams, Greer. 1947. "Laboratory against Death." *Cosmopolitan.* February.
Williams, M.J. 2008. "The Coming Revolution in Foreign Affairs: Rethinking American National Security." *International Affairs* 84(6): 1109–29.
Williams, Raymond. 1974. *Television: Technology and Cultural Form.* London: Fontana-Collins.
———. 1980. *Problems in Materialism and Culture.* London: Verso.
Wills, Christopher. 1998. *Children of Prometheus: The Accelerating Pace of Human Evolution.* Reading, MA: Perseus Books.
Wray, Ricardo J., Steven M. Becker, Neil Henderson, Deborah Glik, Keri Jupka, Sarah Middleton, Carson Henderson, Allison Drury, and Elizabeth W. Mitchell. 2008. "Communicating with the Public about Emerging Health Threats: Lessons from the Pre-event Message Development Project." *American Journal of Public Health* 98(12): 2214–22.
Wright, Susan, and David A. Wallace. 2000. "Varieties of Secrets and Secret Varieties: The Case of Biotechnology." *Politics and the Life Sciences* 19(1): 45.
Wurtz, N., M.P. Grobusch, and D. Raoult. 2014. "Negative Impact of Laws Regarding Biosecurity and Bioterrorism on Real Diseases." *Clinical Microbiology and Infection* 20(6): 507–15.

Young, Donna. 2003. "Drill Tests Seattle's Radionuclide Preparedness." *American Journal of Health-System Pharmacy* 60(13): 1300.

———. 2005. "States, Hospitals Learn Emergency-Preparedness Lessons in TOPOFF 3." *American Journal of Health-System Pharmacy* 62(10): 1000.

Žižek, Slavoj. 2002. *Welcome to the Desert of the Real! Five Essays on September 11 and Related Dates*. London: Verso.

Index

Illustrations are indicated by italicized page numbers.

CPSIA information can be obtained
at www.ICGtesting.com
Printed in the USA
LVOW11s1204151017
552514LV00002B/375/P